THE FUTURE WEALTH OF AMERICA:

BEING A GLANCE AT THE

RESOURCES OF THE UNITED STATES

AND THE

COMMERCIAL AND AGRICULTURAL ADVANTAGES

OF CULTIVATING

TEA, COFFEE, AND INDIGO, THE DATE, MANGO, JACK, LEECHEE, GUAVA, AND ORANGE TREES, ETC.

WITH A

REVIEW OF THE CHINA TRADE.

BY FRANCIS BONYNGE,

FOR FOURTEEN YEARS A RESIDENT IN INDIA AND WEST OF CHINA.

NEW-YORK:
PUBLISHED BY THE AUTHOR.
1852.

Preface.

The unparalleled progress of the United States since the adoption of the Federal Constitution, is perhaps the grandest epoch of all time. The power and grandeur of Rome, reached after centuries of toilsome labor and protracted conflicts, has been more than equalled in the short space of half a century, and the past is but a bright index of the future.

The position of the American Union, not only in the relations of this country to the whole continent, but to the world, is becoming every day more clearly known and defined.

The elements of national wealth and independence are becoming every day of increasing importance, and as an effort to indicate to the people of the American Union additional sources of national and individual wealth, and to point the eye of the country to new and valuable opportunities for developing the national industry, this volume is respectfully offered to the public.

The discussion is divided into three parts. In the first part I have endeavored to present a view of the Cotton and Rice of our country, and to show the necessity of introducing other agricultural staples to meet the increasing wants of America and of the world.

In the second part is contained a hisrory of the Tea trade, instructions in the best modes of cultivating the Tea and Indigo plants, and their manufacture, with a variety of statistical information on the points involved, commercial and agricultural.

A history of the Opium trade is presented in a sub-division of the second part.

In the third, and concluding portion of the work, the author has freely and candidly stated his views in regard to the present and future of America. Having had extensive opportunities of observation in Great Britain, India, China, and the United States, he has been led to discuss the great interests of free and slave labor, as they present themselves to his experience and his convictions; and without arrogating to be right, to the exclusion of others, he defers to the future the correctness of his judgment.

While the author does not wish to introduce the reader to the following pages with an affected apology, he feels called upon to say that should the work be found not to abide the test of literary criticism, he desires it to be understood that it is not offered to the public with such a purpose. Having for fourteen years spent his time in India and China, the acquisition and constant use of Hindostanee or Urdu, the Bengalee and Tartar, French and Italian languages, and the consequent disuse of his mother tongue, have served to prevent him from executing the work, as a literary effort, in a style worthy the magnitude of the interests involved. Yet, trusting that the suggestions and facts presented will be regarded as of more value than mere literary finish, he trusts that they will be attentively considered.

The proceeds of the sale of this work will be used in introducing the proposed staples into this country.

The author returns his grateful acknowledgments to the numerous subscribers for this work who have so cordially entered into the objects it is designed to promote; and, without invidiousness, he may especially mention the names of the following distinguished gentlemen:—J. F. O'Hear, Esq., Secretary South Carolina Agricultural Society; James Rose, Esq., President R. R. Bank, Charleston; Hon. John Schnierle, Mayor, Charleston; Henry Ravenel, Esq., Charleston; Hon. William Aikin, Charleston; R. R. Cuyler, Esq., President R. R. Bank, Savannah; His Excellency Baron Von Jerrott, Prussian Minister, Washington; Dr. Daniel Lee, Patent Office, Washington; Messrs. Tierman & Pringle, Charleston; Edmond Molyneux, Esq., British Consul, Savannah; Messrs. Lobeck & Schepler, New-York.

NEW-YORK, December 1, 1851.

CONTENTS.

TABLE A. POPULATION OF THE UNITED STATES.

	Whites.	Slaves.	Free Colored.	Total.
1830,	10,526,246	2,008,043	320,596	
1840,	14,189,108	2,487,213	386,245	
1850,	19,668,736	3,179,589	419,173	23,267,498
1860,	26,552,793	3,910,894	482,049	30,945,736
1870,	34,518,630	4,810,399	554,356	39,883,385
1880,	43,148,287	5,916,790	637,508	49,702,585
1890,	53,935,358	7,277,651	733,134	61,946,143
1900,	67,419,197	9,051,510	843,104	77,313,811
1910,	84,273,996	11,133,351	969,579	96,376,928
1920,	105,342,495	13,694,029	1,115,015	120,151,539
1930,	131,678,119	16,843,655	1,282,267	149,804,041
1940,	164,597,646	20,717,695	1,474,607	186,789,948
1950,	205,747,060	25,482,764	1,695,798	232,925,622
1960,	257,183,825	31,343,799	1,950,167	290,487,781
1970,	321,479,781	38,553,872	2,242,692	362,276,345
1980,	401,849,726	47,421,262	2,079,095	451,850,064
1990,	502,312,157	58,328,252	2,965,959	563,606,368
2000,	627,890,296	71,743,749	3,410,852	703,044,897

Whites are estimated to increase up to 1860, at the rate of 35 per cent.; up to 1870, 30 per cent.; from 1870, 25 per cent.; Slaves, from 1850, 23 per cent. Free Colored, from 1850, 15 per cent.

Had the Slave been calculated at 28 per cent., which is near their average rate of increase, it would make them, in the year 2000, 128,126,630, more than 1 to every 5 whites.

TABLE B. AREA AND POPULATION.

Free States.	Sq. miles.	Sq. acres.	Free.	Slaves.
Maine,	35,000	22,400,000	582,026	
New-Hampshire,	8,000	5,139,000	318,003	
Massachusetts,	7,250	4,640,000	994,724	
Rhode Island,	1,200	768,000	147,549	
Vermont,	8,000	5,120,000	314,322	
Connecticut,	4,750	3,040,000	370,913	
New-York,	46,000	29,440,000	3,098,818	
New-Jersey,	6,851	4,384,640	489,868	52
Pennsylvania,	47,000	30,080,000	2,341,204	
Ohio,	39,964	25,576,960	1,981,940	
Indiana,	33,809	21,637,760	990,258	
Illinois,	55,405	35,459,200	850,000	
Michigan,	56,243	35,995,520	397,576	
Iowa,	50,914	32,584,960	192,000	
Wisconsin,	53,924	34,511,360	305,596	
Total Free States,	454,340	290,777,600		52

Slave States.	Sq. miles.	Sq. acres.	Free.	Slaves.
Delaware,	2,120	1,356,800	90,277	2,332
Maryland,	11,000	7,040,000	492,661	90,355
Virginia,	61,352	39,265,280	940,000	460,000
North Carolina,	45,500	29,120,000	480,000	280,000
South Carolina,	28,000	17,920,000	280,000	350,000
Georgia,	58,000	37,120,000	555,000	365,000
Kentucky,	37,680	24,115,200	782,000	211,000
Tennessee,	44,000	28,160,000	800,000	250,000
Louisiana,	46,431	29,715,840	250,000	200,000
Mississippi,	47,147	30,174,080	300,000	320,000
Alabama,	50,722	32,462,080	440,000	330,000
Missouri,	67,380	43,123,200	590,000	91,547
Arkansas,	52,108	33,406,720	150,000	45,000
Florida,	59,268	37,931,520	45,000	22,000
Total Slave States,	610,798	390,910,720		
Texas,	325,520	208,332,800	100,000	50,000
Oregon,	341,463	218,536,320		
California,	448,691	287,162,240	200,000	
New-Mexico,	77,387	49,527,680		
Territory N. W. of the Mississippi,	723,248	462,878,720		
Terr. W. Ark. and Missouri, & S. Platte River,	248,851	159,264,640		
Old N. W. Terr. E. Miss. R. and N. Wisconsin,	22,336	636,438,400		
Total Territory,	3,252,684	2,081,717,76		

The total number of inhabitants is 23,267,498, of which the Slaves are 3,179,589, and Free Colored, 419,173.

TABLE C. SLAVE LABOR.

Of one hundred negro children born, there die, say under

1 year, - - - - - ..	21·64
5 " - - - - - -	16·78
10 " - - - - - -	3·79
20 " - - - - - -	7·52
30 " - - - - - -	9·13
40 " - - - - - -	7·94
50 " - - - - - -	8·43
60 " - - - - - -	6·85
70 " - - - - - -	6·20
80 " - - - - - -	4·81
90 " - - - - - -	4·32
100 " - - - - - -	1·76
upwards of 100, - - -	70*

Say 21·64 of the infants who died before one year, lived on an average three months, and had to be supported for that three months; and as many mothers were also taken away from labor at least for three months prior to and after confinement.

Therefore, say one hundred mothers were taken away from manual labor for three months, that would average for one year twenty-five persons, and 21·64 children for three months, for a year, = 5·41. Therefore,

Mothers who did not work and were supported one year,	25
21·64 infants died before one year, - - - -	5·41
16·78 children died before five years, - - - -	83·90
3·79 children died before ten years, - - - -	37·90
57·79 survived ten years and did not work, - -	577 80
7·52 young people died before twenty, averaging, say five years from the ten years already calculated,	37·60
50·26 survived twenty years, and were supported an additional ten years, - - - - - - -	502·60

* From the Census of Charleston.

Of these, 9·13 died before thirty, averaging, say five years from the twentieth year already calculated, -	45·65
41·13 survived thirty years, and were supported an additional ten years, - - - - - - -	411·30
Of these, 7·94 died before forty, averaging, as above, five years, - - - - - - - - - -	39·70
33·19 survived forty, supported an additional ten years,	331·90
Of these, 8·43 died before fifty, averaging, as above, say five years, - - - - - - - - - -	41·15
24·66 survived fifty, supported an additional ten years,	246·66
Of these, 6·85 died before sixty, average support five years, - - - - - - - - -	34·25
17·91 survived sixty, supported an additional ten years,	179·10
Of these, 6·32 died before seventy, average support five years, - - - - - - - - -	31·60
11·59 survived seventy, supported an additional ten years, - - - - - - - - -	115·90
Of these, 4·81 died before eighty, average support five years, - - - - - - - - -	24·05
6·78 survived eighty, supported an additional ten years,	67·80
Of these, 4·32 died before ninety, average support, for them, - - - - - - - - -	21·60
2·46 survived ninety, supported an additional ten years,	24·60
Total number of years' support of the one hundred infants, from birth until the last died, - -	2865·02

Number out of one hundred infants who reached years of maturity, and the work they did from birth to the end of the last remaining one's life:

7·52 died between ten and twenty; these, of course, did some work, say on an average they worked five years in the ten—average worked for one year, -	37·60
50·26 survived twenty years; labored for ten years, say	502·60
9·13 died before thirty, averaging, say five years' work each, - - - - - - - - -	45·65
41·13 survived thirty, and labored for an additional ten years, - - - - - - - - -	411·30

7·94 died before forty, averaging, say five years' work each, - - - - - - - - - -	39·70
33·19 survived forty, and labored for an additional ten years, - - - - - - - - - -	331·90
8·43 died before fifty, averaging, say five years' work each, - - - - - - - - - -	42·15
24·76 survived fifty, and labored for an additional ten years, - - - - - - - - - -	247·60
6·85 died before sixty, averaging, say five years' labor,	34·25
17·91 survived sixty, and labored for an additional ten years, - - - - - - - - - -	179·10
Total number of years' labor performed by the one hundred infants, - - - - - - - -	1876·85
From which deduct one-seventh for Sundays, - -	268·12
Deduct one-tenth for sickness, - - - - - -	187·68
Total years' labor, - - - - - - - - -	1421·05
Number of years' support of the one hundred, throughout their term of life, from birth till the last dropped into the grave, - - - - - - -	2865·02
Number of years' labor performed by the one hundred infants, - - - - - - - -	1421·05
Number of days supported, - - - - -	1,045,732
Number of days worked, - - - - - -	518,683

Or as one, to two, nearly, of days worked to days supported.

Therefore, as one slave costs fifty cents per week, two will cost, - - - - - - - -	$1 00
Life insurance, say $400 each at say three per cent. per annum, - - - - - - - - - -	23
Medical charges per week, say, - - - - - -	7
Total per week, say, - - - - - - - - -	$1 30

Or for six working days, twenty-one cents and two-thirds per diem.

EMIGRATION FROM ENGLAND, ACCORDING TO BRITISH RETURNS.

	North American Colonies.	United States.	Australian Colonies and New Zealand.	All other places.	Total.
1825,	8,741	5,551	485	114	14,891
1826,	12,818	7,063	903	116	20,900
1827,	12,648	14,526	715	114	28,003
1828,	12,084	12,817	1,056	135	26,092
1829,	13,307	15,678	2,016	197	31,198
1830,	30,574	24,887	1,242	204	56,907
1831,	58,067	23,418	1,561	114	83,160
1832,	66,339	32,872	3,733	196	103,140
1833,	28,808	29,109	4,093	517	62,527
1834,	40,060	33,074	2,800	288	76,222
1835,	15,573	26,720	1,860	325	44,478
1836,	34,226	37,774	3,124	293	75,417
1837,	29,884	36,770	5,054	326	72,034
1838,	4,577	14,332	14,021	292	33,222
1839,	12,658	33,536	15,786	227	62,207
1840,	32,293	40,642	15,856	1,958	90,743
1841,	38,164	45,117	32,625	2,786	118,592
1842,	54,123	63,852	8,534	1,835	128,344
1843,	23,518	28,335	3,478	1,881	57,212
1844,	22,924	43,666	2,229	1,873	70,686
1845,	31,803	58,538	830	2,330	93,501
1846,	43,439	82,239	2,347	1,826	129,851
1847,	109,680	142,154	4,947	1,487	258,276
1848,	31,065	214,132	23,904	4,887	248,088
1849,	41,367	219,450	32,091	6,590	299,499
1850,	32,961	223,078	16,037	8,773	280,849
	841,701	1,483,325	201,323	39,684	2,622,617

IMMIGRATION INTO THE UNITED STATES, ACCORDING TO AMERICAN RETURNS.

		1848.		1849.		1850.
Ireland,	-	98,061	-	112,691	-	116,583
Germany,	-	51,973	-	55,700	-	45,404
England,	-	23,062	-	28,321	-	28,131
Scotland,	-	6,080	-	8,400	-	6,771
Wales,	-	-	-	1,782	-	1,520

Total immigration into the States:

1821 to 1831,	-	14,000	1838 to 1842,	-	76,000
1832,	-	45,000	1843,	-	75,000
1833,	-	56,000	1844,	-	74,000
1834,	-	65,000	1845,	-	102,000
1835,	-	53,000	1846,	-	147,000
1836,	-	62,000	1847,	-	234,742
1837,	-	78,000	1848,	-	229,492

AGRICULTURAL AND COMMERCIAL STAPLES

OF AND FOR

AMERICA.

PAST AND PRESENT STATE OF THE COTTON TRADE AND FUTURE PROSPECTS.

MY subject is one of the deepest interest to the United States, both morally and in regard to their material prosperity. I aim at bringing round a reformation, which I shall show to be of the highest importance to this and all other countries. In the following pages, were I to labor to produce conviction, simply by authority of names, however respected, and of probabilities, however plausible, it would be to present the question in such condition as to excite endless disputation, and leave men in doubt. Therefore, I shall state facts, derived from long personal experience, and from historical data; and these facts, derived from those two sources, conjointly shall, I fear not, successfully appeal to the practical good sense of my readers.

In my endeavors to promote the welfare of America, and the cause of humanity all over the world, I will confine myself to the simple relations of actual state of things, to show the pressing necessities of a comprehensive adaptation of other articles of Agriculture and Commerce, for the exercise of the industry and skill of the vastly and rapidly increasing children of America; and to afford to all countries, through their agency, genuine articles, and at comparative low prices—articles that have become necessaries of life, and that will tend to promote not only the prosperity and comfort of all, but promote temperance and sobriety. And, in appealing to every class of American citizens, I will at once pass on to the following, to show the necessity of the introduction of other staples of Agriculture and Commerce than those already existing.

In a country situated as America is, if her people sit down in contentment with the events of yesterday, and with the state of things in times gone by, and will not look forward to the wants of to-morrow, or the future, of themselves and their posterity, adieu to American greatness, adieu to the advancement of civilization within her shores. "Plant a tree, dig a well," says the Shaster (a Bible) of the Hindoos, "for your children's use; these are among the great works of charity." And in like manner are we bound to look beyond the span of our existence, and bequeath to our successors increasing means of happiness, in keeping with the wants of an unparalleled increasing population. Our predecessors provided for us; so, in like manner, it is our duty to do all we can for the rising generation. Or are we to die and leave no other monument to witness our virtues but that cold and too often lying marble slab erected over our miserable clay?

The great staples of exports, cotton and rice, &c., are at a dead stand in respect to the quantity that may be produced, and if not receding, will recede from the present amount produced. In value they have fallen to a very low figure.

Look back to the time prior to the introduction of cotton and rice. What was the condition then of the people? Was it not a general struggle for an humble subsistence? Had not cotton and rice, &c., been introduced, what would be the state of the United States to-day? Would her wealth or her population be to what they now are? These products aided to advance the prosperity and raise these States to great eminence; and sixty years of peace, progress, and agricultural prosperity have placed the United States of America in the proudest position. Will Americans rest content with past achievements? Her staples of production were all-sufficient for the profitable employment of her population heretofore: these staples are, if not falling off in quantity, very much reduced in value; while the population is increasing with extraordinary rapidity, and the extent of land called into cultivation is spread from one sea to the other.

Cotton and rice were golden mines, and all-sufficient for their time, and, left to the support of the few, would go on enriching them to the end of time; but let a great and numerous people place their dependence on them alone and they will become exhausted, and want, idleness and misery must be the consequences.

The present products of America were all-sufficient to the state of things some twenty years ago, but to-day's greatly increased numbers, require something more. And if nothing additional be forthcoming things must retrograde; and America's future be a grievous history.

Look to the progressive increase of the population for the last thirty years, and mark, that the reduction in the value of cotton is commensurate with that increase. For brevity, take only the eight following states, or for the exercise of individual calculation, the table, given at the end hereof of all the states :

	1820.	'30.	'40.	'50.
Georgia, - - -	348,989	516,567	691,392	920,000
South Carolina,	502,741	581,185	594,398	630,000
Alabama, - -	127,901	308,997	590,756	777,000
Mississippi, - -	75,448	136,806	375,641	620.000
Louisiana, - -	153,407	218,575	352,411	450,000
Arkansas, - -	"	30,388	97,574	195,000
Missouri, - -	66,586	140,074	383,762	631,000
Texas, - - -	"	"	"	150,000
Total, 8 States,	1,275,072	1,932,592	3,085,944	4,373,000

Increase from 1820 to '30—51$\frac{1}{2}$ per cent. Yearly, 5$\frac{3}{26}$
" '30 to '40—59$\frac{1}{2}$ " " 6
" '40 to '50—47 " " 4$\frac{7}{10}$

Or in 30 years the population multiplied itself 3$\frac{1}{2}$ times. Placing in view, the progressive increase of cotton for the same period of 30 years.

From 1820 to '30, - - - - -	6,509,587	bales.
" '30 to '40, - - - - -	13,680,004	"
" '40 to '50, - - - - -	21,178,128	"

Increase from 1820 to '30 over preceding 10 years
" " '30 to '40 " " 108[illegible]
" " '40 to '50 " " 54$\frac{10}{13}$

a falling off in progressive increase of 54$\frac{10}{13}$ per cent.,—but probably in the next ten years there will be no increase over the last ten. To ascertain how long the production has been at a stand still, let the above be divided into periods of five years.

1820 to '25— 2,315,998					
'25 to '30— 4,293,589	Increase, $85\frac{3}{8}$ per cent. compared with preceding 5 yrs., yearly 17 per cent.				
'30 to '35— 5.556,485	"	30 per cent.	"	6	"
'35 to '40— 8,123,520	"	$46\frac{1}{5}$ "	"	9	"
'40 to '45—10,122,306	"	$24\frac{5}{8}$ "	"	5	"
'45 to '50—11,052,822	"	$9\frac{1}{5}$ "	"	2 nearly.	

This shows that the produce is nearly stationary for the last ten years. In fact, if the above time had been divided into periods of six years instead of five, it would show as follows:

From 1826 to '32, increase per cent., compared with preceding six years.

'32 to '38	"	45	"	or yearly	$7\frac{1}{2}$	per cent.
'38 to '44	"	$46\frac{5}{14}$	"	"	$7\frac{2}{3}$	"
'44 to '50	"	$7\frac{3}{4}$	" only, or	"	$1\frac{7}{24}$	" only.

So it may be said that the production of cotton in America has not increased materially for twelve years—there has been only an increase of 160,000 bales yearly, for the last six years, over the former six.

If the above time be divided into two periods of twelve years it will be found the price of cotton was follows:

From 1826 to '38	-	$12\frac{1}{2}$ cent. per lb.
'38 to '50	-	$8\frac{14}{15}$ " "

showing a falling off of 29 per cent in value.

Taking the total value in dollars as follows, viz:

From	1820 to '25—2,316,900	bales	a	cent.	per lb.	$
"	'25 to '30—4,293,589	"	a	10	"	$171,743,200
"	'30 to '35—5,556,500	"	a	$11\frac{17}{20}$	"	263,378,000
"	'35 to '40—8,123,500	"	a	$12\frac{17}{30}$	"	408,494,800
"	'40 to '45-10,122,300	"	a	$7\frac{3}{5}$	"	307,717,600
"	'45 to '50-11,053,000	"	a	$8\frac{10}{15}$	"	383,170,400

Since the year '39, the cotton crop is comparatively stationary in quantity; and it is seen that the cotton crop realized more in dollars for five years—from '35 to '40—than it has in the two following periods, that is from '40 to '45, and from '45 to '50. Therefore, if the population bo taken at the period of '35 of the same eight States, it will be found, while the amount of dollars realizable has fallen off in the two last periods from the first in the amounts of $100,723,040 and $25,324,400, that the population has increased from 2,466,528 to 4,373,000—increase, 1,956,000 persons: wherefore, so far as cotton goes, there has been a decline in the last ten years of $126,047,440, compared with the former five; therefore the means that four persons enjoyed from '35 to '40, has been diminished one-sixth by lowness of price, and, on the other hand, they have to divide that diminished means with three other persons; and this decrease is in the UNITED STATES ONE GREAT STAPLE! This is a bad state of things; and young people have now little opening for their energy and abilities: and young men must necessarily rush into every desperate undertaking that holds out to them any dim prospect of support; if not, to pass their best days in the miserable professional look-out for a brief, a patient, or some official employment, to afford them a poor pittance.

Therefore, this all-important staple—cotton—admits of no further immediate increase for the employment of additional hands. Are they then to turn to rice cultivation; that has receded both in quantity and in price already; or to sugar cultivation, that, too, will be found very precarious, and it requires the best lands? To manufacturing,—is it rational to advise some 22,000,000 of people, thinly scattered over the mighty area of 4,375,-000 miles, to turn spinners and weavers, and machine-makers, &c.? What would support them? The legitimate field for American industry is the American soil.

It is said Texas has the best soil in the Union. Let the population of the eight States named, having to day 4,373,000 souls, increase at the rate of 20 per cent., (I put it down at 20 per cent., because, without other staples for cultivation, the Southern States cannot progress; they have run their tether's length in cotton and rice, and labor will yearly become of less value), in every 10 years, their population would be, in 2010, or 160 years hence, 81,856,000 souls. Therefore, if 4,373,000 are obliged to share among them that property in its diminished state, which 2,466,000 enjoyed some 10 to 15 years ago, what will be done with the population when doubled, which, instead of 40 years, may be in 20 years more? While the population of all the States is increasing without parallel in its rapidity, the exports of America are far behind all moderate expectation, and all the promises of antecedents. Is there a superabundant population in America, that an opening must be made for employment by entering into a hard struggle with all the world, and that world's cheap labor? In 160 years hence, the eight States named may have a population of 81,000,000, and, at that distant period, the whole population of Georgia, South Carolina, Alabama, Mississippi, Louisiana, Arkansas, Missouri and Texas, would but sufficiently populate Texas alone, leaving not a soul in the other seven States!

Cotton has done wonders for America; but everything will get old. The cotton States themselves complain that the land yields them nearly one-third less produce, and it is seen it sells for one-third less price. Therefore, let people who are anxious for the onward prosperous course of this dear land of freedom and industry, look around them—let them inquire what is going on in other countries—let them ask themselves, is there no possibility of competitors to arise, and to cope with America? and in making these inquiries, let no American say there is no soil, no climate, no people to equal American soil, climate and people. No man can inquire aright with such feelings. The General who holds his adversary in contempt is most likely to be overcome. I say there is danger, and I say so from actual knowledge of other countries, from my visit to them; and I say distinctly, it is not the want of suitable soil, climate, and cheap labor, that has left the cotton trade so largely in the hands of America.

Look to the East—Smyrna and Egypt are arising from the sleep of ages, and their export of cotton has increased threefold the last three years. Even railroads are now making. The Emperor of Constantinople is exerting himself to the utmost, and this year has been distributing seeds to his subjects. The French are alive at Algiers. In Brazil, cotton cultivation is extending. The West Indies are doing a little. East India is advancing rapidly. Well, America may laugh at the idea of Egypt and Smyrna, or Brazils, or West Indies, &c., &c., monopolizing the cotton trade; but it is well to remember each will do a little, and that little is increasing, and, as the Scotchman would say, "every little makes a muckle."

Shipments to England from the following places in 1850, were—

	Bales.
West Indies, - - -	75,087
Brazils, - - - -	171,322
Smyrna and Egypt, - - -	79,505
America, - - - -	1,182,656
East Indies, - - - -	369,220

Therefore, these countries export more than one-half as much as America to England, and they export also to other countries than England. And, as will be shown, while America is nearly stationary for the last years, that the above countries have gradually increased.

15. The cultivation in America is in its senile years; in the above countries it is in its infancy—that is, in some, and in others in their lately renewed efforts; and they, taken together, are even now formidable rivals, and America will have to struggle hard. But as far as East India is concerned, it would not be the least surprising if she monopolized altogether the cotton markets in a few years hence. That she can do so there is no doubt; that she has not done so under the half-dozen planters who the East India Company have taken out there from Georgia, is no evidence to the contrary. The cause of the failure is one to be accounted for; the planting in the East is below 20 degrees of N. latitude. All Southerns know that it will not even do well low down in Florida, and that the cultivation is northwards of the 27 deg. to 36 deg. N. latitude.

16. It is wished to keep up the price of cotton to 12½ cts. per lb. There is only one way to do it, viz.—to prevail upon Egypt and Smyrna, Brazils and East India, to fall back to that position they held in the cotton market in 1838—to agree amongst your own States to produce

900,000 bales yearly less than you have done for the last 12 years; that is, from 1838 to 1850. Do this, and then you will get $12\frac{1}{2}$ cents per lb.

17. In support of this statement, I offer the following:

From 1827 to 1838 inclusive, 12 years' produce, was
14,048,000 bales, at $12\frac{1}{2}$ cents per lb., $567,890,400
" 1839 to 1850 inclusive, 12 years' produce, was
25,545,000 bales, at $8\frac{14}{15}$ cents per lb., $635,162,000
The over produce for 12 years is
11,497,000 bales, and difference in price, $67,271,700

Giving for that over produce about $1\frac{5}{8}$ cents per lb. Therefore, if America falls back to a yearly average produce of 1,170,000 bales, and the above countries enter into arrangements as aforesaid, she will succeed in the wishes of her planters.

18. Let cotton committees, who rack their brains in order to find a loop hole in nature's laws to effect impossibilities, continue to go on with over production, and keep up the prices forsooth, build sheds to store it away from the purchaser! Would it not be more safely stored away from him at the bottom of the sea, or a million of bales to be burned yearly? It would puzzle the purchaser. But if 3,000,000 of bales are to be produced, as it is supposed there yearly will be, then there will be 1,830,000 to be destroyed some way. I remember when the East India British Company and England was at war with China, enthusiasts laid before the wondering world, that if China's 360,000,000, would only wear a night cap each, it would employ England's 30,000,000 to weave them. But the Chinese, stubborn people, would not wear more red night caps than usual;

and that grand hope has been dissipated. But it may be well asked, why produce at all any quantity that must be kept behind doors from the purchaser? Let any artificial means be resorted to, and it would have the effect of driving purchasers from America, and enlist their best exertions to promote the growth of cotton in other places. Any such artificial means would be suicidal to the best interest of these States; and they who would attempt such would be the best friends to Brazils, East India, and Egypt. Raise the price of cotton from 8 to 12½ cents per lb., and a premium is held out of 4½ cents per lb. to all other countries, for every pound they may produce. There is no breaking through nature's laws; they are like water, depressed in one part they rise up in another.

It is extraordinary that instead of these crude and dangerous schemes there are none to turn their attention to some practical means of opening legitimate fields for the employment of the industry of the people. Why will no one ask what are the articles we import, and make an endeavor to render their country independent of foreign aid? But alas! there are few, if any; and for one inquirer there ever will be a hundred thousand schemers. The one say, if we produce everything for ourselves there will be no commerce! And such a one would advocate the extraordinary roundabout way, that it is better to produce a superfluity of one material, and be obliged to carry it from the place of its growth to the most distant ports of the world; and then to import, from the most distant ports, tea, coffee, sugar, &c.; and thus surcharge these agricultural staples with a whole posse of ship builders, sawyers, carpenters, painters, seamen, agents, and retailers, all of whom could earn a more certain support in being employed in a productive pursuit. All these become a heavy tax on the productive population. There must always be merchants, and the more articles there will be produced, the more merchants will be benefited. But to say we must take silk, tea, coffee, &c., from China, when we can produce them for less than one-fifth the expense of importing them, is folly.

But besides the wisdom of our producing these things ourselves, there is another reason to urge us to do so, viz: to suppress one of the vilest and most fiendish trades that could be invented or suggested by the common enemy of the human race. I mean that foul, accursed opium trade, which is fast demoralizing and depopulating the Eastern world. Americans! in your trading with China you aid in perpetrating that wrong on the human kind. But more of this hereafter.

Cotton Planters' Meeting

AT LEON.

The Report of the Committee states that, "perhaps, no interests in the world are so surrounded with difficulties, or subject to so many disasters, as the cotton planting interests." This very statement will only show the very uncertain and dangerous position of the whole of the States with regard to this staple. A prudent people would not place their all on so uncertain a foundation. Where is there an article that does not fluctuate in prices? Those articles of consumption which a man must have three times a day to support existence, even they, although there can be no mistake in the amount required, vary in prices, from the uncertainty in production. One year the wheat, another the oats or barley, another the potato crops, might fail; there are years some may fail, yet, from the abundant harvest of the others, plenty and cheapness of food would be the consequence. Yet in these articles, while there can be no deviation from the usual quantities consumed, upon the other hand, from the uncertainty of harvests, cause a great deal of fluctuation. With regard to cotton, indigo, wines, &c., on both sides, there ever must be great fluctuations. The consumption as well as the produce is always liable to great changes, and therefore they will always, let what step may be taken, vary greatly in price. The consumption of these articles is very uncertain in all countries, either from depression of the agricultural population from short crops, or of scarcity generally, or from political causes. Let, for instance, India be threatened with a general revolution, the whole of the exports for Calcutta, Bombay, and Madras, would be thrown on the hands of the manufacturers in England. And whether the American cotton crop was a short one or otherwise, the price of cotton would then go down. A famine over India would not only interfere with the usual consumption during its existence, but for some years afterwards. Do what cotton planters may by combination, they never will be able to grapple with the difficulties of the trade. And if their report be carefully gone through, it will be seen upon what false ground they have made it, and propose it for legislation.

Over production of cotton can alone be the cause of permanent low prices. Well, suppose cotton be down this year to six cents ; to save planters from selling at that price, let one-third part, or any such portion of the year's crop be withdrawn from the market as would raise the price up to 10 cents or 12 cents per pound, that one-third portion would remain round to the following crop, and to be added to it. Well, the following crop may be a large one, too, and another one-third of that crop may be withdrawn also ; therefore, as but two-thirds of each of the former years have been sold, there would remain another two-thirds for the third year ; and, therefore, the third year, without any new crop, would have a full crop ; or, supposing the third year's be a failure, i. e., a complete failure, then there is the crop made up for it. Supposing it was at the option of the American planters to carry the above scheme into practice, what could it effect? It could not raise the cotton, in the aggregate, one cent in price. All that it could do would be to equalize the price of one year with the other, raising up the minimum and reducing the maximum. And the weavers or manufacturers should encourage the scheme, because it would have the effect of rendering the price of cotton piece goods on their side equal also; and, instead of their having to hold over the piece goods for a favorable market, at a great loss, by the accumulation of interest and warehouse expense, &c., the cotton planters would take all that additional trouble on themselves.

Therefore, let it be allowed that the American cotton planter be the sole arbitrator of the market, such a scheme would be detrimental to his own prospects. But, is the cotton planter so far ignorant as to suppose that the same quantity of cotton would be consumed at twelve cents per pound as there would be at six cents per pound. Raise the price of cotton to the uniform price of twelve cents and it would be to contract its consumption. It may be supposed cotton must be had at any price: that is a great mistake; flax and hemp, silk, wool, &c., would take its place; cotton shirting and sheeting, cotton vests, cotton drawers, cotton night-caps, cotton socks, would be all superseded by the use of woolen, linen, and silken goods. The grazier with his sheep, the man who rears the silk-worm, the farmer who sows his flax and his hemp, who are all, in a manner, cotton planters, would be recipients of the benefits. China and India would export more silk; Australia, Great Britain, France, more wool;

Ireland, Belgium, Switzerland and Russia would produce greater quantities of flax, and East India and Russia would export more hemp and jute.

The committee reports, " enough is shown by the facts to establish an important point: that the extent of consumption, up to this time, has been controlled by the extent of production; and we must, therefore, look to other causes for the ruinous depression in price to which we have so often submitted." It is true, that the consumption will always be ruled by production; it is a fact, scarcely needing elucidation, produce less cotton and there is less to be consumed, produce more and more will be consumed.* The greater quantity produced the lower the price, and the lower the price of cotton the more it will be economical for wear instead of silk, woolen or linen. Limit the quantity, raise the price, and, as a matter of economy, place it on a par with silks, woolens and linens, and then cotton cloths would go out of use in geometrical proportion as their price be raised compared with the above. And, instead of cotton sheets, towels, &c., we will have linen ones. Instead of muslin dresses, there will be more silk ones, and more silk and

* For instance, "the failure of the American crop in year 1846, as in the very last season (1850), caused a considerable rise in the price of cotton; and it was calculated that, in that year, an advance in price of two pence per pound, required an increased payment by this country of £4,000,000 ($20,000,000). In this year the increase in price has caused many spinners and manufacturers of coarse yarns and heavy goods either to stop their mills or to work but a short time. It has been well ascertained, that, with high prices of the raw material, the present enormous production of cotton manufactures will not, and cannot, be taken off by the markets of the world."—*Manchester Guardian, July 23rd.* 1850—*from Mr. Royle's work.*

woolen stockings, vests, &c. Therefore, when the production of cotton is above the mark, it reduces itself in price; but at a low figure there is room for consumption *ad infinitum*—to the injury of the sheep owner, the silk manufacturer and the flax planter.

The report goes on, viz: "The second point requiring investigation is the capacity of the world for over production. To this your committee concede there cannot be a definite answer given; they incline, however, strongly in opinion that, at fair prices and with proper organization on the part of the American cotton planters, the capacity for over production does not, and never can exist."

There is here a great want of due attention to the trade of the world displayed. To put the question, can Egypt, for instance, beat us out of the market? is but one consideration out of the many. The first and leading question is, can we produce cotton at that low rate as to create an extensive consumption, viz: to throw woolen cloths, as far as they can be substituted for cotton cloths, out of the market, and to keep linens out of use—to keep silk wear within bounds? Upon that point must rest the sole subject of increase of consumption. Then there is another question, is the flax trade increasing? and the same of woolen and of silk may be asked. If so, people must prize these articles, inasmuch as cotton goods are now so cheap. Well, for the last twelve years, cotton has realized not more that $8\frac{14}{15}$ cents per pound or so. It is now sought to bring up the price to $12\frac{1}{2}$ cents per pound; therefore, that will be to enhance the value of cotton goods some 50 per cent.; and, while the linen, woolen and silk goods remain at present low rates, it is proposed to place cotton goods in competition with them, at 50 per cent. dearer than they have been. This, all will admit, would be advantageous to the three former articles, disadvantageous to the sale of the latter.

Then, with regard to the second point, viz: production, it may be asked, who are the growers of the cotton plant, or who may become eventually planters of it?

There are the Southern States of America, ranks first of all nations; then comes the East Indies, Egypt and Smyrna, Brazils, West India and Islands, China, Italy, &c. Well, all these countries produce more or less cotton, and export one-third as much as America; and probably India produces twice the quantity that America does. If cotton be kept at the present low price of eight cents, the inducement for them to extend their cultivation cannot be great, inasmuch as the price is all but ruinous to the Southern States.

The Southern States' produce last six years over the former six years is as 2,049,000, 2,208,000, or increase in six years of 7^3 per cent. only, or yearly $1\frac{7}{24}$ per cent.

The exports from the undermentioned places was in—

		1848.	1849.	1850.	
East India, - -	Bales,	227,572	182.090	309,220	to L'pool only.
Egypt & Smyrna,	„	29,032	72,725	79,505	„
Brazils, - - -	„	100,244	163,149	171,322	„

Showing the increase of 1849 over 1848 on the above three places had been - - - - - 17 per cent
„ „ 1850 over 1849 „ 34 „
„ „ 1850 over 1848 „ $56\frac{11}{12}$ „

This shows a steady and large increase.

If the produce of the Southern States be taken for the same periods, viz.:—

	1848.	1849.	1850.		
Bales,	2,347,634	2,728,596	2,096,706		
Showing increase of 1849 over 1848			nearly	12	per cent.
„ decrease of 1850 under 1849			„	$23\frac{3}{5}$	„
„ decrease of 1850 under 1848			„	32	„

The calculations for 1851 is under that of the crop of '49, and little more than equal to '48.

Therefore, while the Southern States are looking backward, the above countries are rapidly advancing, and that too when the Southern States are nearly reduced to ruin by low prices. Well, then, raise the price of cotton by any scheme to fifty per cent dearer; two agents will then enter the field, whose united forces will be irresistible, and will drive the Southern States into a much worse predicament than they at present stand in, viz., woolen, linen, and silk, will be called in as substitutes for the then expensive cotton cloth; and Egypt, Smyrna, East India, and Brazils, will move with all their powers, being stimulated thereto by an advance of 50 per cent on the present prices of cotton. However, it is seen that present prices are sufficient to make them renew their exertions, and beyond these measures already taken, as stated, for the forwarding the growth of cotton in the above place. It is seen by official announcement, that the American gins are to be introduced into two countries of Asia, two of Europe, and one of Africa. Well, to keep these countries out of the market, is to keep down the price of cotton; to encourage them, is to give 50 per cent more for their produce; and any inconvenience America may put herself to to bring round high prices, will be only for the benefit of foreigners, and to her own ruin. Therefore, if cotton-planting is not profitable, or so little so as not to compensate for all the risks from short crops, or a complete failure, the only way is to give up planting, and plant, or grow something else. Interference with the course of events will drive cotton planting from these States; while allowing matters to right themselves, will, at least, preserve a fair portion of the cotton trade to America.

Again, the committee states—"The largest five years' average production the world has yet furnished is 2,791,000 bales per annum. That, of these, England, France, and the United States require for their consumption from 2,000,000 to 2,200,000 bales, leaving not more than one-fourth of the annual product to supply the balance of the world, with a population probably ten times as large as their own."

Here is another very erroneous view put forward. It is so well known that England are spinners and weavers for all the world, that it is difficult to conceive how the committee could put forward the statement above.

The imports to England are exported again to India, China, the Continent of Europe, all Asia, Australia, Africa, Russia, and to all places; not more, if as much, as one-sixth part of the cotton imported being used in England. The imports to England for 1849 from all countries were 775,470,000 lbs.; the consumption for that year of Great Britain was 136,420,765 lbs. only.

The committee again states—"Having now shown that there has been no over production in the aggregate, and that there is no reasonable probability that there ever will be, your committee will show the effects of irregular production on prices and consumption." It is seen, from 1826 to 1838, the price of cotton was 12½ cents per lb.;

from 1838 to 1850, only 8 14/15 cents per lb. Irregularity of production would necessarily cause fluctuations in the market. An abundant crop one year might cheapen, while a short crop the next might enhance, the prices; but such irregularity, in the long run, could not pull down the price some 29 per cent. permanently if there was not over production. There is no over production for the world. The world would find people to wear twice the quantity of cotton clothing, but the lowness of prices must be tempting to the world for them to do so. For instance, Americans consume 11½ lbs. per head; it has been calculated that the East Indians consume 20 lbs. per head; Great Britain only 4½ lbs. per head. Here is room, if returns be correct, for an increase in the consumption to take place in Great Britain of 7 lbs. per head, or say, some 210,000,000 lbs. But what is to prevail on Great Britain to wear more cotton cloth? Simply, to produce more cotton, and cheapen the price; not to raise it above what it is at present selling. Cheap prices have given a wonderful impetus to the consumption of cotton; and lower it still, it would go on increasing. Yet the present quantity produced is over production for the cotton planter, inasmuch as he sells his cotton for the last twelve years for 29 per cent. less than the former twelve years. Not only has the prices of cotton receded some 29 per cent., but the productive powers of the cotton land is seriously diminished. Under these circumstances, is it safe to persevere in extending cotton cultivation?

But now, to take another view of the cotton trade of America, in conjunction with that of other countries. China cannot produce sufficient of all things for herself; her soil is covered over with mountains, which take away most materially from her area of arable land, and although the Chinaman holds the first rank of all nations for his ingenuity as a cultivator, and the government of China the first rank as patrons of agriculture, and of the industrious poor man; yet, if the area of land in cultivation, as stated by Gutzlaff, and repeated by Martin, from official returns, be correct, as being but 141,119,347 of English acres, and the population to be 367,632,907 souls. It is only to be wondered how they can export as many articles as they do, especially silk and tea. Therefore China may be a mart always for the sale of cotton, and perhaps it may not be an improper deviation from the subject to hint at the desirability of considering what might be the result of opening a grain trade from the west of America with China. I believe, already American rice has been shipped from the eastern ports to China.

Smyrna and Egypt, in their cotton movement, will be but gradual. However, railroads, and the present awakening spirit, stimulated by exertions of the English, may advance their cotton cultivations, and an advance on the present prices, would be the greatest auxiliary to promote the wishes of the present rulers, and of the English.

However, to pass over the Brazils and West Indies, and consider what are the prospects of East India. In the first place, it is necessary to start with the knowledge that there are two parties in England, to all outside appearance, exerting and straining every nerve to promote the cultivation in East India; while in fact, both parties are doing little or nothing beyond casting blame on each other. Individuals, upon whose exertions alone the growth of the cotton plant would be extended, are not in the field. A few American planters are there upon salaries, and a few servants of the government get up experimental farms; for every man who will try his hand at cotton growing, is sure to be enrolled in parliamentary blue books, and be lionized for the time being; therefore many lovers of popularity have their little to do with cotton planting; and among the whole body there is not one interested person, whose hopes are concentrated, and whose means of livelihood is to be realized from his successful issue in cotton planting. It is lamentable, but it is not less the fact, that people now-a-days carry on all their experimental cultivations, &c., not so much for the purpose of doing good, but to figure in print, and appear to be a *somebody*. When we look at the voluminous reports on cotton cultivation in East India, when we listen to the worthless wrangling of the Court of Directors of the British East India Company, and the Manchester Chamber of Commerce, it must be grievous to find that two such bodies should waste their energies in pen and ink, and high-winded speeches. Every planter of indigo, &c., in East India, is at the complete mercy of a boyish magistrate, or collector; the planter, whatever his position may be, is at their discretion. I can say, after fourteen years in the East Indies, that the serf of Russia has far more security than planters or cultivators have in the British possessions in India. On the other hand, instead of the Manchester Chamber of Commerce or the commercial community employing active and intelligent men—and investing their capital to enable such to make themselves a provision, they keep aloof, talk and write. Latterly, i. e. 1850, the Chamber sent out a commission of inquiry to ascertain the cause why *they did not* GET MORE COTTON FROM INDIA. Had they distributed the money necessary to defray that expensive mission to half a score of active planters, they would have done something towards the object desired.

These are the obstacles, and there are no others to prevent cotton being grown to any extent, and of every quality in the East Indies.

The East Indian cotton contains 25 per cent of waste and dirt, while the American contains only 12½ per cent. It is more difficult to spin the Indian, from its often breaking, and requiring more turns of the spindle, from its shortness of fibre, than that of America. But the yarn made from a pound of East Indian cotton, which cost 3½ pence sterling will sell for 7 pence, while from the American, which cost 4½ pence the lb., the yarn sells for 7¾ pence only, making up somewhat of the difference in value of the yarn.

The capabilities of India to produce cotton is seen from the following evidence. Mr. Chapman, manager of the Great Peninsular Railway Company, stated to Mr. Bazley, "That cotton of good quality for English use is always to be had in Berar at about 1½ pence per lb., ranging of late years from 1¼ to 1⅞ pence per lb. The quality of this cotton is such, that at a certain relative difference of price (averaging about 15 per cent. less for Indian than American), it can be used instead of American for more than 50 per cent. of our manufactures; that is, will afford material for all yarns under No. 20." (Royle on cotton in India.) Berar is part of Central India, three hundred to four hundred miles from Bombay; the railroads now making will open a communication with Berar for exporting that cotton to the coasts. There is little effort made to obtain the best cotton from India, or even the middling qualities in a clean state. Mr. Mercer and Mr. Finnie, American planters, state "that an inferior and dirty cotton is the most profitable article to the native trader, and even to the European merchants and Mr. Petrie, before the cotton committe stated— "Cotton that would be sold in England at 4½ pence per lb., and cotton that would only bring 3 pence, will sell in India within 3 to 4 per cent of the same value. The cultivators know this, and therefore they have no object in bestowing more care and labor." Mr. Read, commissioner of Benares division, wrote 16th August, 1848: "In this division, the cultivation of cotton is little more than nominal, it is most commonly mixed up with other crops, it is nowhere carefully attended to; in many places it is an object of superstitious aversion, and often when the plant contrives to struggle to maturity it is left, after being stripped of a portion of the bolls, enough to supply domestic purposes, to be devoured by cattle." Evidence to this effect might be produced ad infinitum.

Mr. Bruce, in a letter dated 15th March, 1848, says: If the associations and manufacturers who have been memorializing and soliciting the authorities in England, regarding the increase of cultivation of cotton in India, actually wish for it, and will guarantee that all that may be produced through my exertions in this country will be taken by them, and paid for here, and will send out responsible agents to receive charge of the cotton, either here or at Calcutta, I will engage and undertake to produce for them as much real good marketable cotton as they may require, and not cost them, when landed in England, more than about 3½ pence per lb., which I think very moderate, considering the Indian cotton generally realizes from 4½ to 5½ pence per lb. in England.

"The Doab, as well as Bundlecund, has always grown considerable quantities of cotton, and will be able, as we hope to show, to grow it of a better quality as soon as the Ganges Canal, that most magnificant of works for irrigation, is completed."—ROYLE.

"The cotton of Jaloun and Jhansi (writes Mr. Bell) was formerly much celebrated. Koonch is now the great mart in that direction, and it is surrounded by the Jaloon Territory. Mr. Bell, after making careful enquiries, ascertained that the cotton of Baugchenee was that which at present is most in repute with the natives of those parts, who gave about 7½ per cent. more for it than for the other best kinds. The district is on the south of the Chumbul, near Dholpore, and therefore probably very similar in soil and climate to that of Jaloun and Jhansi, which are not very far distant. Mr. Bell further thought, that it might be landed at Liverpool in a clean and desirable condition for about 3½ pence per lb. As these districts, as well as Bundlecund, lie to the westward of the Jumna, and have always been famous among the natives for their cotton, it is probably owing to some physical, pecularities of soil, or of climate, that the cotton is produced of a better quality, for we cannot discover that it has the advantage of any more careful culture or cleaning."—ROYLE.

Some, indeed, ascribe the whole of the difficulties and the unimproved state of the Indian cotton to the continued purchases by the European merchants of the very dirty and adulterated article. These, however, who are resident at Bombay rebut this charge, and explain the peculiarities of their position, by showing the difficulty there is for merchants resident in the capital to come in contact with the cultivator in the country; for they are a small body, not exceeding forty in number, belonging to about twenty firms, so that each firm has only on an average, two resident members, a number barely sufficient for the transaction of local business, and they are moreover, in most cases, the agents of others, whose orders they must comply with—and the execution of these orders is always limited to time, leaving them wholly dependent on the cotton to be found at Bombay, whatever be its quality."—ROYLE.

I will give but a few extracts more to show, even under the indifferent management of parties who were rather pushed to make experiments, than to enter into them with a hearty good will. The Manchester people, who are all anxiety, have done actually nothing beyond scolding the East India Company. Under such state of things, a kind of Punch and Judy struggle, what could be expected? Yet, a few examples of success can be produced, abundantly sufficient to show, that East India will become one of the greatest cotton countries in the world.

Mr. Mercer, the American planter, says, "Dharwar is more like the Mississippi climate than any other he has met with in India. The elevation modifies the climate, which also feels the influence of both monsoons or rains, so that it never becomes extremely dry, and is never inundated with excessive rains." Mr. Mercer finds the seed returning to its original Mexican character, instead of deteriorating as in other parts of India. Here the natives, witnessing the success of cultivation on the government farm with American cotton, have themselves adopted and are rapidly increasing it, as they find it yields a larger quantity per acre, and they get a better price, even from the native weavers."

The culture of American cotton has been greatly increasing in quantity; from 600 acres the first year (1842,) to 15,000 acres this year, 1845, Mr. Mercer thought that not less than 1,000,000 lbs. of American cotton would be produced. This crop has now been ginned, and 40,000 rupees (20,000 dollars) worth, at 66¼ Rupees per 784 lbs. (i.e. 33 dollars in round numbers) has been bought from the ryots (farmers), for transmission to this country."

The further expense of this cotton amounted to 17½ rupees, (exchange value of a rupee is 1s. 10d. or say 45 cts.) Some of this cotton (100 bales) was sold at Bombay for 113 rupees a candy, and the 500 which were sent to Manchester for 6½ pence a pouud, having cost 3¾ pence a pound.

Mr. Blount wrote from Dharwar, the following two letters to Mr. Royle, viz.—" I have much pleasure in informing you, that the number of acres planted this year is in round numbers, 15,400 against 3,400 in 1848, and we have had a copious monsoon, and the expectation is, there will be another good productive season; should such be the event, I have no doubt the crop will double itself another year."—*Dhawar, Oct.* 27, 1849.

" The fine crop of New Orleans, likely to be realized this season, will, I think, induce a more extended cultivation of that variety; I should not be surprised to see the cultivation come up next year to 30,000 acres. I have now completed the despatch of cotton purchased in

1849 for government, and can give you a correct quantity.—New Orleans, 330 candies, whieh will cost, put down to Liverpool, $3\frac{78}{100}$ pence per lb.; native cotton 60 candies, cost, put down at Liverpool, about $3\frac{1}{2}$ pence per lb. The cotton is superior in quality to the previous crop."—*Dharwar, Aug. 8th,* 1850.

"Since October last my partners purchased, on account of my firm here, a parcel of New Orleans cotton, which is now on its way to this country; and, by the mail, arrived this evening. I have advice of a further purchase on account, of an eminent Lancashire spinner. The quality of this cotton is now well known, and appreciated in Bombay, and its value is quoted in all the prices current; on the 15th January, the quotations were 114 rupees, whilst the highest quotations for Surats were 74 rupees. The return to the cultivator, upon an acre of ground, is thus doubled; for he gets 50 per cent. more weight of cotton, and an increase of 50 per cent. in value. I feel, therefore, most confident, that this will give a stimulus to the native grower, and that we shall see a rapid increase in the production, as well as a great improvement in the quality."—Signed, JOHN PEEL.

I give these few extracts from Mr. Royle's work, (in which is to be found a collection of all that has transpired on the cultivation of cotton in India; printed in London this year,) simply to show that the American cotton planter has some cause to apprehend a severe competition, and to suggest the great danger there may be in tampering with the natural order of things, because, should planters by any means advance the price of cotton from present rates, they will give an impetus to cotton cultivation all over the world, that must be detrimental to America.

I have made quotations for which I feel I may guarantee, not only their accuracy, but that they also fairly represent the state of cotton cultivation in India, where, I have no hesitation to say, after fourteen years residence, and after a tour through South Carolina and Georgia, that India has every means of producing as fair cotton, and far cheaper than America.

And I submit, under the circumstances, the necessity of America to introduce some other staples to uphold the interest of agriculture and commerce than cotton. In the first place, because the trade now in cotton is in a bad state. Second, that cotton is not sufficient in itself, even in the most prosperous state, to employ the increasing population of America. And in the third place, England is determined to render herself independent of America, and is most likely to succeed.

It is curious, however, to remark that the East India Government has been making their experiments in India over to North West, under the influence of a burning hot wind, or South of N., Lat. 20°. Had the experiments been made N. East, from Lat. 24° to 29°, a climate and soil like that of South America might be obtained.

Review of the Rice Trade.

I WILL enter into a very brief statement of this, the second staple of the Southern States. It is the one most dangerous to the life of the planters.

I will take the last 24 years of the returns embracing North and South Carolina and Georgia, and the last 14 years for review of the price of the article.

		Tierces.		*Tierces.*
From 1825 to '30,	six year's produce	676,816,	average	112,802
" 1831 to '36,	" " "	661,311,	"	126,885
" 1837 to '42,	" " "	648,458,	"	108,076
" 1843 to '48,	" " "	774,988,	"	129,164

Which shows a decline, comparing the last 12 years with the former, of 14,681 tierces, or more than one per cent.

Shipments to Europe.

Taking the last 24 years.

From 1825 to '36—12 years, . .	668,669 tierces.
" 1837 to '48 " . .	566,216 "
Decline in the shipping trade, .	102,423 tierces.

or yearly 8,535 tierces.

Taking the prices from 1835 inclusive, to 1848—14

years, and placing the prices of the first 7 years in juxtaposition with the last 7, the result would be as follows :

Prices from 1835 to '41, 7 years averaged $3 30 to 3 88 pr. tierce.
" " 1842 to '48, " " 2 94 to 3 57 "

Showing a falling off in price of 10 per cent. Or if if the two last years be taken, the prices of which I obtained after making out the above table, it would be as follows compared with prices from '34 to '41 inclusive :

Prices from 1835 to '41 averaged yearly $3 30 to 3 88 per tierce.
" " 1849 to '50 " " 2 83 to 3 31 "
or nearly 15 per cent.—or, taking the last nine years compared with the preceding 7, viz:
Prices from 1835 to '41, 7 years averaged $3 30 to 3 88 pr. tierce.
" " 1842 to '50, 9 " " 2 88 to 3 44 "

Showing a decline on the nine years of 12½ to 13 per cent. in prices. This decline concerns South Carolina most; Georgia grows but ¼ as much as S. Carolina, and North Carolina only produces about 7,500 tierces yearly.

The rice planters for the latter portion of the last nine years, had the most favorable opportunity to obtain a high price, compared with former years. The general failure of the potatoe crop over Europe for a series of years, and a short crop of grain on the Continent of Europe; famine among eight millions of people, and great scarcity in the highlands of Scotland; all, have not been sufficient to keep up the prices of American rice.

There must be a cause for this falling off. Scarcity must be supplied. Indian corn had been shipped to Ireland, but people have every disrelish to it. Corn has not made up the supply.

The only causes to be found for the falling off in the shipments and in price, (as it has not arisen from over production), are the imports of rice and paddy from Arracan, Patna, and Benares. In Arracan, taking the bulk, the Arracan rice is some 500 per cent. cheaper than American rice in their different countries.

The Arracan sells in England for 10s. or \$2½ the cwt.
" American " " " 18 or 4½ "

Therefore it is, although the American brings 80 per cent. better prices, and freight must also be lower, that the Arracan rice is ousting the American.

Why the East Indians are able to undersell the American planter, will be readily perceived from the following, which I have already stated in the Charleston Courier, &c., viz: the rice planter in India, with his wife and family, labor in the field. The economy of the people is striking. The man's clothing consists of a strip of coarse cotton that passes between his legs, one end attached before, the other end behind, to a string that surrounds his waist. Two pieces of such clothing, of the best description, will last a couple of years; and all his year's clothing will not cost 50 cents. The diet of the people is rice, (which sometimes they use in raw state, simply steeped in a little water), which, when cooked, they eat with some few cooked vegetables mixed up with pepper, chilies, huldee (ginger), and clarified butter. Sometimes, instead of vegetables they have fish, and if a goat be killed in the village, then there may be flesh instead of vegetables and fish. A man may support himself for 50 cents a month, which is about the value of one rupee of their money. However, this specimen of simplicity and nakedness, it may be supposed, could not enter into competition with his brother rice planters, the Hon. this one and that of S. Carolina and Georgia, who would not relish wine less than thirty to forty years old, and who live up to that in all things else, and are to be found the great lions of the seasons at London, Paris and Rome; who, possessing from 100 to 1000 slaves, each to act at his beck, with the precision of a well-disciplined company of infantry. Yet so it is; the simple poor Indian, from the circumstance of his position, is too much for the civilized lord and master of the many. Look at the ant; it can do nothing of itself; but look at the ants, they can by their united powers raise up mounds fifty feet high, which to look at would be supposed to be caused by some convulsions of nature; yet, dig into it, and it would be found to be the home of these little mites—the structure of their persevering efforts. If the people of Arracan or Bengal be considered, they may be compared to the family of ants. Their great numbers enable them to bring under rice cultivation, not a farm, not a town's land, but a whole district, as far as the eye could penetrate; yea, some several score miles in extent. And the climate and nature of the country is such, that little more is required than casting the seed on the ground. The embankments inclosing the little fields of so many, is rendered comparatively inexpensive, not being of near the same extent as if each plantation had its own embankment. Throughout Bengal government takes care of the bunds (embankments).

But is there no other cause to fear than foreign competition? It may be said there is a boundary in America to rice cultivation from limitation of suitable land. Let the rice planters consider well!! Do they not at present, at least some of them, find in their possession a little piece of machinery at which they rejoice exceedingly. It throws up water with the greatest rapidity, and irrigates rice estates in the twinkling of an eye. Now, if that engine or any other remedy be found to give command over the course of the waters flow, and if irrigation can be so effected, whole districts may be called into that cultivation; and then, who will purchase a rice plantation at the rate of one hundred dollars, more or less, per acre? yea, who would purchase at all when they could get land for a song? Let rice planters dwell upon this point, and reflect if rice be fifteen per cent. lower now than it has been nine years ago. Should double the quantity of land be brought into cultivation what would be the consequence?. The first effect would be from the use of such power over water, that there would be no sales for rice lands beyond the value of the formerly waste lands. The second would be, the influx into the cultivation would reduce prices to a rate that would leave rice land valueless and the rice planters beggars. The rice planter, as well as cotton planter, and every American, is deeply interested in promoting the cultivation of other articles, to divert the attention of the public from the two only channels now left them, and keep them from being choked up by the dense mass of people who must, from hard necessity, rush into them.

I believe the foregoing review of the existing position of these States suffices to show the necessity of enterprise.—the necessity of their people doing something practical to relieve the two afore mentioned articles of commerce from depression. To give some opening for the employment of at least a portion of the cultivators; for, if all go on cultivating cotton or rice, all must be ruined.

LIST OF ARTICLES PROPOSED FOR INTRODUCTION INTO AMERICA.

I beg to lay before the American community the following list of articles for introduction into the country; and, before proceeding to do so, it may be right to say, that, with the subject I am about to enter on in the following pages, I am perfectly conversant, from a long experience of fourteen years; that my knowledge of the tea plant, Indigo plant, and these manufactures, the date, the mangoe and leeche tree, &c., &c., is not derived from others, or are the following pages a compilation from magazines, &c. I have been five years cultivating and manufacturing indigo, saltpeter, &c., &c., and nearly six in the cultivation and manufacture of tea, &c.; and ten years in the field, at one time or the other, interested in every department of agriculture, has given me an acquaintance with the subjects that I may say, without fear of contradiction, no other person has had. My visit to America was in consequence of the advice and encouragement of the Hon. Abbott Lawrence, your minister at St. James.* Prior to my correspondence with Mr. Lawrence, I had some communication with the consul at Dublin; both gentlemen laid the communications before the Washington Executive Government, and when Mr. Lawrence received a reply, he advised me to come over here personally.

* See Correspondence in Patent Office Report of 1851.

I will now beg to insert copies of letters from several gentlemen known to the citizens of America generally:

AGRICULTURAL ROOM,
U. S. Patent Office, May 5th, 1851.

The bearer, Mr. Frank Bonynge, is a gentleman who has had considerable experience in the culture of tea and indigo plants, and their manufacture in the East Indies, and visits this country for the purpose of establishing a tea plantation in such locality as shall appear most favorable. I have examined an essay from his pen of thirty-five pages, seen his letters of commendation from Liverpool, and, from information derived from Baron Von Gerott, Prussian minister resident in Washington, I believe him to be a man of integrity, and his purposes an object of great importance to the Southern States.

(Signed) DANIEL LEE,
And Editor of the Southern Cultivator,
AUGUSTA, GA.

And, in another letter, Dr. Lee further writes, "Your communication to the Hon. Abbot Lawrence, on the culture and manufacture of tea, will be published in my next official report, of which Congress has ordered 130,000 volumes printed and bound."

Yours respectfully,
(Signed) DANIEL LEE.

F. Bonynge, Esq.

I have conversed with Mr. Bonynge, seen a letter of introduction from Dr. Lee, of Washington, and read a manuscript of Mr. Bonynge's on the subject of the tea plant, and also with the productions of some plants, fruits, &c., as well as the indigo. I have read a letter from Daniel Willis, of Liverpool, recommending Mr Bonynge to Messrs. Loison and Silvester, of New Orleans, speaking in most favorable terms of him. . . I have no hesitation in stating, that Mr. Bonynge possesses more information on this subject than any person with whom I have conversed.

(Signed) JNO. BACKMAN.

CHARLESTON, S. CAROLINA,
Mayor's Office, July 18th, 1851.

This will certify, that Mr. Francis Bonynge has some time passed been zealously engaged in bringing to the notice of the citizens of Charleston the advantages of introducing into the South Atlantic States of Carolina, Georgia, Florida, &c., the tea and indigo plants, the date, mangoe, and leeche trees, &c., &c.

A considerable interest has been awakened in this city, among a respectable portion of our citizens, on the subject thus introduced by Mr. Bonynge, and there is a fair prospect of future encouragement to him in his undertaking.

He is about leaving this city on a temporary visit to our neighbouring cities and towns, to submit his proposals to such as may be disposed to contribute their aid, by subscribing to the above object. Mr. Bonynge has made a universally favorable impression on all who have made his acquaintance here, and warrants me in recommending him as a gentleman, intelligent and experienced in the department he is engaged in, and who, by his approved intercourse among us, is entitled to the encouragement and favorable consideration of every well-disposed citizen or gentleman.

(Signed) JOHN SEHNIERLE, *Mayor.*

MAYOR'S OFFICE.
Savannah, 28th July, 1851.

I cheerfully concur in the foregoing recommendation of Mr. Bonynge to the encouragement and favorable notice of our citizens, &c.

(Signed) ROBERT H. GRIFFEN,
Acting-Mayor.

SAVANNAH, JULY 12TH, 1851.

MR. FRANCIS BONYNGE,
Dear Sir,

Your favor of the 10th instant is at hand. When you first arrived, I was on the eve of departure for Washington, but before I left there was at my request a favorable notice in the Savannah Republic. I left for you a letter to Dr. James Scriven, a gentleman of wealth and affluence, and a thorough planter. I am but just returned home, and have not been able to see the planters of the vicinity, but I think many would subscribe.

I own two tracts of land in the south-western part of the State—one of 250 acres in Early county, and one of 490 acres in Irwin county, either or both of which I will cheerfully contribute. I shall be here for several days, and would advise your visiting Savannah.

Very respectfully,
Your obedient servant,
(Signed) R. R. CUYLER.

F. BONYNGE, ESQ.,
My dear Sir,

I have just seen an announcement of your arrival in Charleston. It seems your object is to introduce among us several plants and trees hitherto untried, and but little known here. I would be pleased to see such experiments fully carried out. Having in South Carolina every variety of soil, I am inclined to the belief, that some, if not all you propose, would succeed. Under this impression, I offer you a location on my lands near this place. I have on my property (the C. and R. R. runs through my plantations, about twenty-six miles above Columbia) every variety of soil particular in this section of the State, also two other varieties some distance from the road. You can have your choice, after examination. &c., &c.

Yours, with respect,
(Signed) OSMOND WOODWARD.

Taken from the three Daily Papers of Charleston.

"At a meeting of the Agricultural Society of South Carolina, held at their farm on Tuesday, the 15th inst., the enterprise of Mr. Bonynge (late from India), to introduce the tea culture, as well as that of the coffee, the date, mangoe, indigo, and other tropical plants, into the State, was brought to the notice of the society. Mr. Bonynge being present, the society came to the resolution, that the enterprise was a laudable one, and calculated to advance the prosperity of the country. They highly approve of the views of Dr. Lee, of the Patent Office department, Washington, as published in the daily papers of our city recently, and tended to Mr. Bonynge the use of their farm on which to commence his experiments."

(Signed) JOS. F. O'HEAR, *Secretary.*

From the Southern Cultivator.

TEA CULTURE IN THE SOUTHERN STATES.

THERE is a gentleman in Georgia who has had several year's experience in India in the culture of tea plants, and the manufacture of tea; and it seems to be our duty to bring the matter fairly before our readers. For reasons, not of a personal or private character, we have for some time abstained from making public the information in our possession on this subject. As early as July, 1850, the Hon. Abbot Lawrence sent a communication from Mr. Francis Bonynge (the gentleman in Georgia to whom we allude), addressed to the Secretary of the Interior, on the production of tea in this country, which came to the Agricultural Desk in the Patent Office, occupied by the writer. We have deemed the letter of Mr. Lawrence, and the communication of Mr. Bonynge, of sufficient interest to insert in the Agricultural Report for 1850; and we have read a paper of some thirty-five manuscript pages, written by Mr. B., to be used before the British Parliament, and seen him several times during his stay in Washington.

If success in the growing and curing of tea leaves depended on the very precarious life of an insect, like that of the silkworm, we should be extremely cautious how we encouraged even an experiment in the business. But the simple matter of picking the green leaves, is the great labor in tea-making, as picking is the serious work in cotton culture. From twenty to thirty pounds of green leaves are a day's work for a feeble Asiatic to gather; and we have good reason to believe that a hundred hands in China do not, on an average, pick so much cotton in a day, as fifty in the Southern states. Indeed, this is the secret, or one of them, why the Chinese cannot, and do not, compete successfully in cotton culture with the readers of this journal. A tea tree needs to be replanted only once in twelve or fifteen years; and an acre will yield about I200 pounds of green leaves a year, which will make 300 pounds of merchantable tea. Mr. Bonynge employed some two hundred hands, and manufactured tea, after the leaves were gathered, at less than an English penny per pound. The East India Tea Company is now making about 200,000 chests a year, and produce a very superior article. The people of the United States annually consume over 20,000,000 lbs.; and those of Great Britain over 50,000,000 lbs. It is truly one of the greatest staples of civilized man, and one that we regard as coming legitimately within the sphere of Southern climate, soil, labor, capital and enterprise. Of course, we esteem it as a matter of experiment only; but an experiment which ought to be fairly made, for if successful, incalculable advantages to the South will certainly follow.

We want that Mr. B. should see the tea plants near our friend Mr. A. R. Kilpatrick of Trinity, La., referred to by him in the May number of the Cultivator. The trouble of procuring any considerable quantity of the tea seeds that will grow after they arrive in this country, is quite a drawback to the enterprize. The Patent Office has received some bushels, but not a seed that vegetated. The operation will be better conducted in future; at least we hope it may. We have before us an interesting communication from Mr. Williams, American Consul at Canton, on the introduction of China fruits into the

United States. Mr. Bonynge has spent fourteen years in the East, and describes a variety of coffee acclimated in a region so high above the ocean, that the tree bears well in a climate subject to pretty severe frosts. It should be borne in mind that cotton itself is a tropical tree—not naturally an annual plant, as we cultivate it in the region of frost. There is nothing improbable in our finding coffee trees that will flourish in all our Gulf States, as far North as 150 miles from the coast. Mangoes and other fruits are also worthy of trial. A chest of tea has been brought from Shanghae to the White House, for the President, in sixty-five days, via San Francisco and Panama. Once it took nearly three years to circumnavigate the globe; now, with good luck in meeting steamers, one can go round the world in 140 days. Indeed, sailing vessels have come from China to California in thirty-three or thirty-five days. A man must be slow if he cannot live a century in the next thirty-three years.

From the Charleston Courier.

TEA CULTURE, ETC., IN THE SOUTHERN STATES.

We received, yesterday, a visit from Mr. Francis Bonynge, a gentleman who has spent fourteen years in the East, actively engaged in the cultivation and manufacture of indigo, sugar, saltpetre, tea and coffee, and whose present object is to introduce into the Southern States the culture of the Tea plant, the Mango tree, Date tree, Coffee plant, &c., and the melons and vegetables of the East Indies, and to carry out the manufacture of the tea leaf, and also of the indigo plant, and to give a full and fair trial to both tea and indigo.

Mr. Bonynge informs us that the soil and climate of

the Southern States are more suited to the cultivation of tea, than those even of China; and that indigo, which was, by-the-by, formerly produced here, can be grown to any extent; and that the coffee plant, in all probability, would flourish here to great advantage, inasmuch as the soil and undulating nature of the land would be in its favor, and the cold of the latitude of this city is not so intense by thirteen degrees as that of the east of China. In fact, Mr. Bonynge has seen this plant growing wild in N. latitude 27° 30″, on hills of from three to five hundred feet in height, where, too, there was an abundance of frost, snow, and hail.

Our space will not allow us at present to give further particulars of this matter; Mr. Bonynge has with him the strongest testimonials in favor of his project from the Hon. Abbot Lawrence, our Minister at the court of St. James, Daniel Lee, Esq., of the Patent Office in Washington, and editor of the *Southern Cultivaior*, and other gentlemen alike distinguished for their position in society, and their literary and scientific attainments, which he will take much pleasure in showing to those who may feel desirous of becoming fully acquainted with the subject. We ourselves regard the introduction of these plants into our State a great *desideratum*, and consequently call the attention of our planters, and such of our citizens as may be interested in the matter, to the visits of Mr. Bonynge to our city.

CULTIVATION OF TEA, INDIGO, ETC., IN GEORGIA AND FLORIDA.

We are indebted to a friend for a most interesting correspondence, some of the details of which we hasten to lay before our readers. Mr. Frank Bonynge, now in Charleston, will soon visit Savannah, with a prospectus for furnishing tea, indigo, and other East India plants, which are calculated to grow in Georgia and Florida. Mr. Bonynge has the best possible evidences that he may be relied upon. He has passed fourteen years in the country where these plants grow, and is perfectly acquainted with the whole subject. His essay on the culture and preparation of tea, &c., will form an important part of the next Patent Office Report.

Subscribers to this important undertaking will have a claim to twelve tea plants, twelve mangoe plants, twelve datetree plants, twelve leechee tree plants, twelve coffee plants, four ounces of melon seed, each kind, half pound indigo seed (if required). Subscriptions are $50 each. Subscribers to the amount of $100 will be entitled to the above, and any other plants from India which they may desire. $25 will procure one half of the above quantity of plants.

We are persuaded that this enterprize of Mr. Bonynge is destined to be a source of vast profit to the Southern States, through the agency of slave labor. It only remains for a few gentlemen in Georgia and Florida, by their subscriptions, to do an immense probable benefit to their respective States. We commend this project to our planters in serious earnestness, and we recommend Mr. Bonynge to the friendly reception of our fellow citizens, when he shall arrive here. That tea can be grown successfully in Carolina, Georgia, and Florida, is almost certain, because the experiment has been pretty fairly tried. The thermometer at *Shanghai* indicates a cold more severe by 13° than in Charleston, S. C. The cold winter of 34, 35, which destroyed the oranges on Mr. Middleton's plantation, left his tea plants uninjured. Mr. Bonynge has seen coffee growing wild in North latitude 27° 30′′, on hills of from three to five hundred feet in height, where, too, there was an abundance of frost, snow, and hail. As for indigo, that substance has already been grown to great advantage in this State.

We proceed to give some interesting statistics from a communication of Mr Bonynge.

From the Baltimore Sun.

TEA IN SOUTH CAROLINA.

The Charleston Courier notices the arrival in that city of Francis Bonynge, a gentleman who has spent fourteen years in the East, actively engaged in the cultivation and manufacture of indigo, sugar, saltpetre, tea, and coffee, and whose present object is to introduce into the Southern States the culture of the tea plant, the mangoe tree, date tree, coffee plants, &c., and the melons and vegetables of the East Indies, and to carry out the manufacture of the tea leaf, and also of the indigo plant, and to give a full and fair trial to both tea and indigo.

Mr. Bonynge says that the soil and climate of the Southern States are more suited to the cultivation of tea than those even in China, and that indigo, which was, by-the-bye, formerly produced in the Southern States, can be grown to any extent, and that the coffee plant, in all probability, would flourish there to great advantage, inasmuch as the soil and undulating nature of the land would be in its favor, and the cold of the latitude of Charleston, is not so intense by thirteen degrees, as that of the east of China. In fact, Mr. Bonynge has seen this plant growing wild in N. latitude 27 deg. 30 min., on hills of from three to five hundred feet in height, where, too, there was an abundance of frost, snow, and hail.

From the Augusta Sentinel.

TEA CULTURE.

We take pleasure in laying before our readers the annexed Prospectus proposing to introduce the tea culture into the South.

Mr. Bonynge has spent fourteen years in India and China, where the tea and other plants, which he proposes to introduce, are cultivated, and professes to be thoroughly acquainted with their cultivation. He comes well and favorably commended by men of high character, and we take pleasure in commending him and his enterprise to the favorable consideration of the people.

PROSPECTUS

For the introduction into America of the tea plants, mangoe, date and leechee trees, the indigo and coffee plants, and all the various kinds of table vegetables, yams, &c.

Subscribers to the undertaking will have a right to

12 Tea Plants, 12 Date Plants,
12 Mangoe " 12 Leechee "
12 Coffee " 4 oz. of Melon seeds of each kind.
½ lb. E. I. Indigo Seed.

Subscribers of $100 will be put in possession of the above, and any other kind of plants they may desire, a list being supplied by Mr Bonynge.

Subscribers to this amount will be entitled to have an agent instructed in tea culture and manufacture, and in steeping, vat beating, precipitating, and boiling, &c. of indigo.

Subscribers of $50, will be entitled to the above plants and seeds.

Subscribers of $25 will be entitled to one-half the above plants and seeds.

Gentlemen upon whom Mr. B. cannot personally call will please to direct to the care of R. R. Cuyler, Esq., or J. P. Screven, M. D.

FRANCIS BONYNGE.

GENTLEMEN WHO HAVE SUBSCRIBED FOR PLANTS IN ORDER TO FORWARD THE INTRODUCTION OF TEA, INDIGO PLANTS, THE DATE, LEECHEE AND MANGOE.

Hon. Wm. Aiken,	Charleston,	Planter,	$100 00
T. B. Lucas, Esq.	"	" and Merch.	"
Dr. Jas. Moultrie, Esq.	"	Private citizen,	"
J. R. Mathews, Esq.	"	Planter,	"
Wm. Middleton, Esq.	"	"	"
J. E. Carew, Esq.	Merch. & prop. of the Mercury		"
James Rose, Esq.	Pres. R. R. Bank & Planter		"
Charles Alton, Esq,	Charleston,	Planter	"
John Ravenel, Esq.	"	Planter & Merch.	"
Henry Ravenel, Esq.	"	Pres. Union Bank,	"
Wm. C. Dukes, Esq.	"	Planter & Merch.	"
E. Vanderhost, Esq.	"	"	"
Edw. Barnwell, Esq. Jr.	"	" and Merch.	"
Charles Macbeth, Esq.	"	" and Lawyer	"
Rev. P. A Lynch, D. D.	"		"

South Carolina Agricultural Society, their farm at Charleston.

R. R. Cuyler, Esq., Savannah, Geo., President R. R. Bank, 750 acres.

J. P. Scriven, Esq. M. D., Planter,		$100 00
Hon. Langdon Cheves,	"	50 00
Wm. P. Bowen, Esq.	"	50 00
W. W. Starkie, Esq.	"	50 00
John Wm. Anderson, Esq.	" and Merchant	50 00
John Williamson, Esq. Merchant		25 00
Rt. Rev. Stephen Elliott, D. D.		25 00
James Hamilton Cowper, Planter		50 00
Dr. R. Moore,		
N. C. Trowbridge,		25 00
D. L. Clinch, Planter		100 00
A. B. Lawton, "		25 00

TEA PLANT.

CULTIVATION, STATISTICS, TRADE, &c.

SOIL AND CLIMATE

OF

AMERICA AND CHINA

COMPARED.

It has been supposed that the Chinese could bring their barren mountains and hills under tea cultivation. They may do so, but they never can make barren or sterile mountains, nor any unfavorable soil produce productive tea trees. It is a physical impossibility for the tea plant to be productive in other than a soil that would be capable of producing other things. Tea trees as well as all plants, trees and vegetables require nourishment; and the richer and deeper the soil the better.

The plant likes a loose loamy soil, of a yellow to a reddish color. It does not like a hard stiff earth, nor will it do at all, in a dry parched or baked earth; sand with clay mixed, if deep, would do well; or a clay soil with sand of two and one-half to three feet deep would do well also.

The root of the tea-tree penetrates the soil downwards in pursuit of sustenance; therefore if the soil be not very rich, but deep, it will do well; and for the same reason the tea plant can support a very severe frost, as its root extends below its influence. The root penetrates directly downwards, having none of any size extending horizontally on, or near, the surface of the earth, for frost to injure. In case of a very soft March, and moisture that might force out the young leaves, and then frost coming on in April, it would injure the young leaves, not the tree, and that crop might be lost, but the other three would be all safe.* However, the leaves of the finer teas are collected soon after budding, and might therefore escape; and in any case it could be only a part of the young leaves that could be injured from an irregular night's frost. It happens in China, that the April crop is at times more or less damaged by frost. However, I do not not refer to a slight hoar frost, but a smart night's freezing.

I have taken up a tea tree of some 35 feet high; there were but a very few horizontal weak roots. The mean, or tap root, was near three-and-a-half feet in length.

Mr. Ball had been told by the Chinese, that a vast improvement was effected in green tea, by bringing the plants from the hills into the plains, and by cultivation and manure, and that this practice had existed for 600 years. The Catholic missionaries stated to Mr. Ball, that "the soil should consist of vegetable mould, sprinkled with sand, light and loose, and rather moist;" and, again, the missionaries replied, "that the tea plant may be planted either in a rich or poor soil, sandy or garden soil, but that which is moist is most suitable;" and, again, they add, "Garden grounds, and the embankments of gardens or fields, are the most favorable." It may be seen from

* Four crops yearly.

this, that what tea requires is depth of soil and moisture. "It is planted as a hedge-shrub, both in China and Japan, and along the ridges of the fields."—Catholic Missionaries. The soil of Chusan is very light and sandy; tea is grown on it for domestic use, not for export. The soil of Amoy and Quang-Tong is a stiff hard soil, unfit for tea; it is grown in both districts, but is of so inferior a quality, and there being no possibility to roll the leaf, which is hard and dry, and the returns from it are so trifling, that the natives will not manufacture it beyond the simple drying of the leaves, which they take into Canton in baskets for sale, and dispose of it at two to four cents per lb. This tea is very largely mixed with the good teas, and sold to England and America.

It is of these trees the Chinese give us seeds. Amoy is in the 24th deg. N. lat., and Quang-Tong is the province of the city of Canton, N. lat. 23 degs.; and these plants being at hand, the Chinese give us as much seed as we require. The proper tea seeds are to be found some 1,000 to 1,200 miles from Canton, and 260 miles from Shanghae. Traveling is slow work in China; there are no steamers or railroads there, and that part up from the 25th deg. of N. lat. is exceedingly mountainous. To get good seed, is not to be accomplished. The East Indian British Government (and no party had the same opportunities), could not succeed; and brought round to Calcutta large quantities of these seeds, which they sent to the North-West Kamoun, and to the North-East Assam. I cultivated, and had several thousand plants; the Assam Company too, the only other party cultivating tea, got in proportion. Neither the Company nor I could get any tea from them, except we stripped off all the leaves, and, like the Chinese, dried them, in which state we could find no market.

TEMPERATURE.

Tea will not bear great excess of temperature. It would live in the open air in England or Ireland, if it had time to take root. There is a large plant in Kew Gardens exposed, but I believe they give it some kind of protection in the cold season. I am of opinion that, with perseverance, tea might grow in Ireland, as fancy hedges, by a great deal of care being taken the first year. But these pots in which all plants, at least all exotics, are condemned to linger out a few years of a miserable, stunted existence, are the bane of all success. Look at a tea plant, with its length of root, in an eight or nine inch pot!

There is no country subject to greater extremes of temperature than China. In the month of April, even as low as 31 deg. of N. lat., the cotton and other annual plants are frequently destroyed by frost; and in the 29th deg. of N. lat., the tea plants are obliged to be covered over with rice straw, and bound with ropes, to protect them from frost and snow. Mr. Fortune says he saw the thermometer stand as high as 100 degs., in the shade, at Shanghae, and that it invariably fell to 12 degs., Farenheit, in winter.

It is, in a great measure, the great difference in the altitudes of China, that causes so great varieties in its climates. Fogan, which is in 27 degs. 4 min. N. lat., is mild throughout, as the following will show:—

			°	″
Fogan,	-	Mean of the four hottest months,	82	0
"	-	Annual mean, - - - -	67	0

		°	″	
Fogan, -	Winter mean, - - - - -	57	0	
Charleston,	32 deg. 45 min., N. latitude.			
"	The mean of the four hottest months,	81	34	average.
"	Annual mean, - - - - -	66	45	"
Savannah,	32 deg. 5 min.			
"	Annual mean, - - - - -	68	3	"

Fogan is the southern point of the tea district, which stretches northward up to 31 degs., and among the high and cold mountainous regions. Probably, in the interior of China, in valleys, it may be grown some degrees higher; and it is not to be doubted that it is grown higher for domestic use.

If the comparisons of the climates of America and China be followed up, it will be seen that the temperatures of 32 degs. 5 min. and 32 degs. 45 min. of Savannah and Charleston are the same as 27 degs. 4 min. of Fogan, China. The highest altitude of Georgia, America, is under 2,000 feet; the latitudes, from 27 degs. upwards, in China are of a very great altitude, being mountainous regions. The cold in America will be very much moderated when cleared of its continuous forests.

EXPENSE OF CULTIVATING TEA IN THE VALLEY OF ASSAM, AND IN THE TARTAR COUNTRY, AND RETURNS.

The following estimate of expenses of cultivation and manufacture of tea on 1,000 acres, was made out for a member of Parliament in 1850:

One Superintendent, $250 per month, yearly, -	$3,000
" 1st Class Assistant, $125 " " -	1,500
" 2nd " " $75 " " -	900

Clearing, and transplanting, and keeping clear first year 1,000 acres, - - - - -	4,875
Elephants, horses, &c., purchase. - - -	600
Building tools, &c., - - - - - -	2,000
Total, - - - - - -	$12,875

SECOND YEAR.

European Superintendence, &c., - - -	$7,200
Weeding, hoeing, and native head establishment,	2,375
Manufacturing, say 40 lbs. tea to an acre, or on 1,000 acres 40,000 lbs. at, - - - -	2,500
Tea-chest and packing charges, $2 per 80 lbs, -	1,000
Total, - - - - - -	$13,075

RETURNS.

100,000 lbs. at say 33⅓, - - - - -	$33,333

THIRD YEAR.

European establishment, - - -	$8,700
Hoeing and weeding 1,000 acres, - -	2,350
Manufacturing, say 200 lbs. per acre, of tea on 1,000 acres=200,000 lbs. at $3 for 80 lbs., -	7,500
Chest-packing, charges to Calcutta, - -	5,000
Total, - - - - -	$23,550

RETURNS.

200,000 lbs. at 33⅓ cents per lb., - -	$66,666

FOURTH YEAR.

European and head establishment, - - -	$9,000
Hoeing and weeding 1,000 acres, - - -	2,350
Manufacturing, say 360 lbs per acre, $1 75 cents per 80 lbs., - - - - - - - -	7,500
Chests, packing, and charges to Calcutta, -	9,000
Total, - - - - - -	$27,850

RETURNS.

360 lbs. per acre, on 1,000=360,000 lbs. at 33⅓ per cent., - - - *l* - - $120,000

The 360,000 lbs. of tea would cost by the estimate, 27,850 dollars, or 7¾ cents per lb. for cultivation, manufacture, and transportation, some 1,400 miles, to Calcutta.

As the above items are condensed, it may be necessary to explain them. Salaries of Superintendents on the fourth year were to be—Superintendent, 350 dollars; 1st Assistant, 250 dollars; 2nd Assistant, 125 dollars a month, salary; making 8,700 dollars, leaving 300 dollars for repairs of their houses, &c. Hoeing and weeding—Hoeing the land in the cold half of the year, and weeding it once during the rains.

To hoe an acre of land in America 2 men are allowed, and of course it would not require more to cut down young soft grass, therefore there would be 4 men to each acre: for Indian labor there are allowed 20 men at 6 cents per diem each: would make, on 1,000 acres, 1200 dollars only. Then, in addition, there is allowed for native head establishment, support of elephants and horses, &c., and a very liberal margin for the managers to save to the company.

Manufacturing item is made out at one man to pick 20 lbs. green leaf, wages 6 cents, and one man to manufacture 40 lbs. into dried tea—7,300 dollars.

Chest and packing, &c., all the wooden and leaden part of the tea-chest would be sent from London to Calcutta, and from Calcutta to Assam, and consequently be very expensive.

Under all these difficulties, which would not exist in America, the tea on the fourth year would not stand in more than $7\frac{3}{4}$ cents per lb., of which 2 cents might be saved in packages and transmission to Calcutta, making the expenses, at the very outside, only $5\frac{3}{4}$ cents per lb.

ACTUAL EXPENSES ON 45 ACRES OF TEA LAND IN THE TARTAR COUNTRY.

Item	Amount
My own table per month, $50 - - -	
Hoeing once, and weeding once, 45 acres. 10 men to hoe, and 10 men to weed an acre. Wages, 6 cents, for hoeing and weeding 45 acres, - - - - - - - - -	54
Picking leaves, one man 20 lbs. of green leaf—one acre produced (average) 128 lbs.—on 45 acres 5,760 lbs., at 6 cents for every 20 lbs. -	172
Manutacturing, one man to every 40 lbs. green leaf—5,760 lbs., at 6 cents for 40 lbs., -	86
Chest, per 80 lbs., (quantity put into a chest) 50 cents, chest for 14,400 lbs. dried tea. These chests were second-hand China tea-chests, - - - - - - - -	90
* Transit to Calcutta, 25 cents per 80 lbs., -	45
Firewood and Charcoal, 7 cents per 80 lbs for 14, 000 lbs., - - - - - - -	12
Horses, two—Elephants, two—10 dollars per month.	
or yearly, 120 dollars, / Table expenses 600 " } $720 to be divided on 2,000 acres, or say $\frac{1}{40}$ part	
Total, - - - - - - -	$459

Return, 320 lbs. the acre on 45 acres—14,400 lbs.† sold at 63 cents. wholesale, - - - $9,072

* Of former year's teas.

† This tea, with a large quantity of other tea, was burned by the Tartars; the teas of the season prior, sold for 63 cents.

14,400 lbs. of tea cost, all expenses included, say 477 dollars, or 3¼ cents per lb. when it would be landed in Calcutta.

The following year, in the middle of the season, I was attacked by the Tartars, and the above tea, which, from want of boats, I could not ship to Calcutta, and all the tea then manufactured, and in process of manufacturing, was destroyed.

PROBABLE EXPENSE OF THE CULTIVATION OF THE TEA PLANT, AND MANUFACTURE OF TEA IN AMERICA, ON 100 ACRES.

I will not insert an item for superintendence, as the planter would know best the care and expense necessary to manage 100 acres of land under cultivation. The price of seeds cannot be ascertained; nor is that of any import as tea trees last 25 to 30 years; so the planter would have to purchase seed but once.

The seeds are sown the first year in beds, from which the young plants are transplanted out into the land intended for plantation. I suppose one-half acre of seedlings would answer for 100 acres of a plantation.

I have given the expense of the first year, on the estimate of 1,000 acres of land, at 13,075 dollars. Circumstances called for that heavy outlay. The company was supposed to be formed in London; the scene of operation near the west of China, under management of agents; and therefore, however little the work the first year, the establishment must be full. In an unhealthy climate, as is that of Assam, it would not be prudent to trust to one superintendent, as the natives will not work but when closely watched; and in case of sickness, there must be some person to take charge. The case is different in America, where every farmer may have his tea plantation attached to his house.

The expense the first year would be for hoeing, pulverizing, drilling, &c., half an acre of land, and sowing the tea seeds, say six men. I have calculated that every slave stands his owner in 21 cents a day. I will give the calculations at the end of the book.

Weekly, say one man one day for ten months, 43+6=49 days, at 20 cents, say, - -	$10

SECOND YEAR.

The transplanting might be done in October, November, January, and February, or as the weather would permit. Say the "tea year" is from 1st September to end of August.

Clearing underwood, say four men per acre, or 100 acres at 80 cents per acre, - -	$80 00
Hoeing 100 acres, four men per acre, at 20 cents each, or 100 acres at 80 cents per acre,	80 00
Transplanting 100 acres, two men per acre, or 40 cents per acre, - - - - - -	40 00
Hoeing the earth round plants, two men per acre, 100 acres, - - - - - -	40 00
Plucking 160 lbs. of green leaf per acre, or on 100 acres 16,000 lbs., one man to pluck 30 lbs.; therefore, 30 lbs. would cost 20 cents, or 16,000 - - - - - - -	106 60
Manufacturing, in the absence of machinery, one man to every 60 lbs. of green leaf= 16,000 lbs., - - - - - - -	53 30
Packing cases for 80 lbs., say 50 cents, for 4,000 lbs. manufactured tea, - - -	25 00
Sieves, of cane or bamboo—(I cannot say what they would cost here, but making in India	

would cost about five cents)—say 50 cents each, 30 - - - - - - - -	15 00
Firewood and charcoal, say 10 cents for every 100 lbs., or 4,000 lbs. - - - - -	4 00
Cast-iron pans, four, at 2 dols. each, - -	8 00
Total expenses, - - - - -	$451 90

RETURNS.

40 lbs. per acre, or on 100 acres 4,000 lbs., which may be valued at 100 cents for some few years, - - - - - - -	$4,000 00

THIRD YEAR.

Hoeing or ploughing 100 acres of land, say -	$40 00
Plucking leaves—800 lbs. of green leaf per acre, one man to pluck 50 lbs., or 20 cents for every 50 lbs.—80,000 lbs., - - -	320 00
Manufacturing, in absence of machinery—one man to 60 lbs. green leaf, or 20 cents for every 60 lbs.—80,000 lbs., - - - -	266 60
Sieves (additional, 100), at 50 cents each, -	50 00
Pans (4 pans additional), at 2 dols. each, -	8 00
Charcoal and firewood, - - - - -	20 00
Packing cases, containing 80 lbs., 50 cents, 20,000 lbs. dried tea, - - - - -	125 00
Total expense, - - - - -	$829 60

RETURNS.

Per acre, 200 lbs. of dried tea, or on 100 acres 20,000 lbs.; say at the retail prices, or a little higher first years of introduction of tea into America, and afterwards 20 cents per lb., - - - - - - - -	$4,000 00

FOURTH YEAR.

Hoeing over 100 acres, - - - - -	$40 00

Sieves (additional), 50, at 50 cents each, -	25 00
Plucking leaves—say 1,200 lbs. of green leaf per acre—on 100 acres, 120,000 lbs., at 60 lbs. for 20 cents, - - - - - -	400 00
Manufacturing 60 lbs. green leaf at 20 cents, 120,000 lbs., - - - - - -	400 00
Charcoal and firewood, 10 cents per 100 lbs. dried tea, or on 30,000 lbs. manufactured, -	30 00
Packages for 80 lbs. 50 cents, on 30,000 lbs. -	187 50
Total expense, - - - -	$1,082 50

RETURNS.

Per acre, 300 lbs., or 30,000 lbs. on 100 acres; say valued at 20 cents per lb., - - -	$6,000 00

It is necessary to say a few words on some of the above items:—

Second year	I gave	4 men to clear brushwood, per acre.
"	"	4 men to hoe the land, per acre.
"	"	2 men to plant it, per acre.
"	"	2 men for a second hoeing, per acre.
Total,	-	12 men, per acre.

I believe planters will say that that is sufficient. There is not much underwood in the American forests, compared to the forests of the north-east of India, where even four East Indians would clear an acre of brushwood.

After the second year, the land, if hoed once, would be sufficient, except there was a very rapid growth of grass or weeds; even the clearing away of such would be but the smallest decimal, not $\frac{1}{10}$ of a cent per lb.

The plucking of leaves, I put down 60 lbs. for one man. This is more than I used to get picked for me in the eastern frontier of India, but it has been done several times. I gave a certain amount for every 20 lbs. of green tea leaf, 8 pice, or about $6\frac{1}{4}$ cents; for 40 lbs. double that sum; and for 60 lbs. treble of it. Men, women, boys, and girls, each picked 20 lbs. from 8 o'clock A. M. to 2 o'clock P. M. Many of the men, and some of the women, plucked 40 lbs., and there were some who would go out early in the morning, and collect 60 lbs. by 6 o'clock in the evening. The work I gave men to hoe, was one acre to ten men. Well, if two men can do that work in America, I should say, one man would pick 60 lbs. of green leaf in the day; for if one man equalled five Indians at hoeing, I do not see why he could not equal the best Indian picker of leaves, for in both works it depended on the arms and perseverance; picking leaves does not require the same physical powers as hoeing; it is more of application that is needed. And it is in that quality that Easterns are defective.

The manufacturing is easy, except that in the coarser kinds of teas it is difficult to roll the old leaves. One man can easily work as much as one will pick. I had, for making charcoal, cutting firewood, and manufacturing, three men to every five pickers of leaves. With machinery to roll the leaves two men's labor would be economized.

In the fourth year, on the 100 acres, I have shown that 30,000 lbs. of tea cost to manufacture, &c., $1082 50, being but 3³ cents per lb., and that, by use of machinery, the quantity might be manufactured for $2\frac{2}{3}$ cents per lb., calculating the expense of slave-labor at 20 cents per diem.

But calculating it at 50 cents per day, for free labor, it would be as follows:

Item		Amount
Hoeing 100 acres, 2 men per acre	-	$100 00
Sieves, (additional) 50 at 50 cents each	-	25 00
Plucking leaves, say 1200 lbs. per acre, of green leaf, one man 60 lbs. at 50 cents, or 120,000 lbs.		1,000 00
Manufacturing, one man to 60 lbs. green leaf, or 120,000 lbs. -	-	1,000 00
Charcoal and firewood, 10 cents per 100 lbs. dried leaf, or 30,000 lbs. -	-	30 00
Packages for 80 lbs., 50 cents, or on 30,000 lbs.		187 50
Total expense	-	$2,342 50

30,000 lbs. for $2,343, or $7\frac{4}{5}$ cents per lb.

This would be a means of not only enriching the cultivator, but of keeping up the price of labor to some $180 a year, and would leave the cotton trade and rice trade to fewer hands. It would give employment to the many, encourage immigration, and give to all a greater degree of prosperity.

The trade of a tea maker might be an item the first season, but after the first crop every man in the business would be *au fait*. Therefore I do not put down any item for a tea maker. The rolling of the leaves might be done by machinery, and would, at the first estimate, in which I allow the expense of a slave at 20 cents a day, be a saving of $1\frac{1}{3}$ cents per lb.; or in the second estimate, wherein I allow the hire of labor to be 50 cents per day, a saving of $3\frac{1}{3}$ cents per lb. Leaving the cost of labor 50 cents, still with machinery, simple in its structure, and therefore of very trifling cost, tea would cost the planter only $4\frac{7}{15}$ cents per lb.

It is most desirable that these estimates be keenly scanned, and to enable all to do so, it will be well to explain the item of produce, so that nought may be taken on *ipse dixit.*

It is shown that on the second year there would be per acre, from 1,210 young plants, 160 lbs. of green leaf collected throughout the year, which would be equal to $2\frac{1}{8}$ ounces for each tree. Now, if that $2\frac{1}{8}$ ounces be divided by four, and any party take a young peach plant, for instance, and on the leaves coming out in April, pluck such as would not injure the tree; it will be found there would be but a few leaves required to make one-half or even one ounce. But entering on the fourth year, the acre is made to produce 1,200 lbs. green leaf, or 1 lb. of green leaf from each tree during four crops. A good tea tree will grow up to a height of 30 to 35 feet; but for facility of plucking leaves it is kept to seven, eight, or nine feet in height. Any person so disposed may select a peach tree of that height, of most luxurious foliage, and ascertain, if he cannot get 1 lb. of leaves off it, if not in one collection, at least in four, made throughout a year.

The tea tree is an evergreen, and its foliage is so rich that the eye cannot penetrate through it; birds, or anything of that description, could rest within its branches without it being possible to observe them. The tree throws out white blossoms of sweet fragrance, and when the tree is plucked it throws them out irregularly. There is nothing so delightful or refreshing as the fragrance from a tea house when manufacturing is going on.

A tea plantation once raised is a permanent property; it would not answer those rolling-stone habits of the modern Scythian of North America, who wanders from farm to farm, and from State to State. A tea plantation may go on for 100 years. A tree will last 25 years. When it shows any symptoms of decline, it only requires to drop a seed by its side. No labor, no care, no loss of time is thereby occurred. If a garden hedge be required, a fence round a field, or ornamental hedges, it will answer all such purposes, while, at the same time, it would yield a most valuable crop.

Let it be inquired, if it is a newly-heard-of thing, that tea can be produced under some six cents per lb.?—or if it be a fresh piece of information, having no other grounds for its truthfulness than the well-arranged figures of estimates?

EXPENSE OF CULTIVATION AND MANUFACTURE OF TEA IN CHINA.

It is known that all China teas sell on an average for 20 cents per lb. at Canton, after crossing mountains and valleys for 1,000 to 1,200 miles, and passing through some six parties' hands, each making his profits; and if it be considered, that the tea that cost at Canton 20 cents per lb., by the time it is retailed here averages 100 cents per lb. it will be seen the price is enhanced five times. Well, the transportation of tea from the districts in China to Canton is more expensive than is its passage from Canton to America, and it passes through on the China side twice as many hands as it does on this side of Canton; therefore, if the price of production be only on the other side of Canton enhanced in proportion as it has been on this side, it would leave the cost of cultivating, manufacturing, &c., at four cents per lb.

The writer of these papers has shown, that in the Tartar country his teas cost him, landed in Calcutta, $2\frac{1}{3}$ cents. On Selling plantation, under the care of an assistant, Mr. Locke, he made tea at 5 rupees for 80 lbs.,* or about $3\frac{1}{8}$ cents per lb.; on Ballijan plantation, in charge of another assistant, Mr. Peters, for 5 rupees 8 annas per 80 lbs., or $3\frac{7}{16}$ cents per lb. Or, if the reader will refer to Java, he will find that tea had been cultivated there under the very illiberal patronage of the Dutch government, who lent funds to the planter, from whom the government received the crops, and allowed for expense of labor, manufacture, and profit for the planter, only $2\frac{1}{2}$ pence sterling, or 5 cents, per lb. This is sufficient to show that the cost of production varies from 2 to 5 cents per lb.

STATE OF THE PEOPLE IN ASSAM, AND THE SINGPHOO, OR TARTAR COUNTRY, WHERE THE ABOVE TEA HAD BEEN CULTIVATED.

If this article seem to be a digression, it is for the purpose of comparing the opportunities America has over the above countries.

The Assam country is the north-east frontier of the British East India Company's territory, from 25 degs. to 28 degs. N. lat., and 93 degs. to 95 degs. 30 min. east long. of Greenwich. The Singphoo, or Tartar country, lies over between that and west of China. The Assamees are, perhaps, of all people, sunk to the lowest state of degradation to which human nature can be lowered. Opium, that accursed drug, has depopulated one of the richest valleys that can be found. The tra-

* Rupee = 50c. A mond is the eastern measure, and contain 80 lbs.

veller may now wander through that very picturesque vale, surrounded on all sides with lofty mountains of the Himmaleh range, varying from 100 feet upwards to 20,000 feet, and inclosed in an amphitheatre of these mountains, with their snow-white tops breaking on the view, as the clouds cleared away, or dropped down their steep sides. There rolls the mighty Burampooter at his feet—being the connected streams of the Dihon, Dibong, from north west and north-east, and the Dihing from the east, China—extending some ten to twenty miles in width, rushing down in its angry course, sweeping huge trees along, the roar of its waters re-echoing from hill to hill, and the wild woods on all side; no town or village or dwelling visible, no human being or boat to be seen,—all strikes awe into the heart of the visitor; and be he a free-thinker or atheist, the grandeur, the wildness of the scene, yea, the tumultuous rage of that great river itself, would force from him the acknowledgment of a Supreme Power. It is in that valley that Providence seemed to show forth most majestically his wonderful works; on that valley he seemed to take a peculiar pleasure to pour all natural blessing for the residence of man; and in that valley man has cursed his own existence, and by the vileness of his nature reduced himself to a state of distress not to be conceived, through the use of opium. The tourist, after long and painful travelling through the dense forest, following wild elephant tracks, the only opening in the woods, sometimes going back from the place of his destination, and sometimes north, sometimes west, as the break made in the underwood might lead, with leeches crawling up his boots, and making their way all over his body, or dropping down on him from the trees, and the blood from their bites, from himself and his followers, trailing in their course. Unexpectedly, he debouches on a miserable village; perhaps there may be ten or fifteen huts, or there may be more, with the greater portion uninhabited or in ruins. The people he finds unintelligent; they stare at him in wonder; one-third drawing huge limbs after them, affected with elephantises; no cultivation visible; but the forest threatening to enclose them in its dark folds.

The man, if there be an infant in question, is the nurse; the mother takes the man's place, and does all the labor, and when that is done turns nurse, and the man resorts to his opium cup, or his opium pipe (for opium is drank mixed in water, or smoked as tobacco from a pipe), and then casts himself down on a piece of tree-bark, or a mat, on the damp floor, where he contracts all his diseases. Infants are few. Opium destroys the powers of procreation; and the greater portion of Assamees go without a child to a premature grave. In the village may be found, perhaps, a score or two of women; not more, probably, than some four or five grown-up men. There was a village near the Koojoo plantation with only two men and eighteen women. Opium is used in the Tartar country as well as Assam. During life no man is so miserable as the opium-eater. In a few years he becomes a thin, emaciated, miserable wretch, incapable of exertions; and when left any time without the use of that fearful drug, he rolls himself on the ground, in the most miserable plight, crying piteously, "Kane ne! kane ne!" There is no opium.

It is among such people that tea in Assam, and the Singphoo, or Tartar country, has been cultivated; and with such workmen the writer has made teas from two cents up to four cents per lb.

THE SINGPHOOS.

The Singphoos, or Tartars, who came originally from Japan, possess an extraordinary degree of energy, and suffer not so much from opium; however, they do not give themselves up to the habitual use of it. In the interior of the country all is disorder; chief is against chief, and clan against clan. It is the same on the western frontier, bordering Assam. But on the borders of Assam they are given to plunder; and prior to the British taking Assam, they used to rush down on her, and sweep away her people and cattle, and, man and beast alike, they sold to the people in the interior; and so the Singphoos, being in possession of slaves, thought it slavery to work. Consequently, there never was enough of provisions to last for the whole year, and when their supply would be exhausted, they then sought in the woods for yams, the soft top of the bamboo, and other vegetables of a similar kind; or they would watch from behind a tree a whole day to get a pot shot at a pig or a deer, for they had but little powder, and their ball was a hammered piece of iron, and they could not afford to lose either by missing fire. Elephants they shot with a poisoned dart, fired from a gun. The poison was a root called *bee*, which they pounded, and put on the dart; the dart was twelve to fourteen inches long. The tusks they took away and sold; the body was useless. Their religion was blood for blood. I heard of a young man who killed a neighbor in a fight. He had to flee from the murdered man's relatives; but two of them followed him several months until they found him out at Suddea, watched him at night until they found where he concealed himself to sleep, and fell upon him and killed him and burned his corpse, and next morning displayed his ashes in a bamboo, as a testimony to show they had performed their duty, and revenged the spirit of their departed relative. Such were the characteristics of the two classes of people I had to deal with to make tea.

Would Americans have no advantage over such people? If tea could be produced for two cents a pound by their agency, why cannot the people of these United States do so likewise?

CULTIVATION OF THE TEA PLANT IN DIFFERENT COUNTRIES, VIZ.:

In China,
" Assam, Eastern Frontier, E. India,
" Tartar country, joining W. of China,
" Kamoun, N. West of E. India,
" Java.

China tea has already been fully discussed.

ASSAM—PLANTERS.

THE Hon. East India Co. had a nursery as a school, to teach the people the art of tea making.

The Assam Tea Co., with a capital of $2,500,000, had several plantations; and Mr. Francis Bonynge had four plantations.

These were the only parties who entered into the cultivation in Assam. The Assam Co. is still in existence. A few of the officers, who derive salaries from the concern, are allowed to carry it on. The funds they have on hand were not of any importance to divide. Their failure arose from the state of Assam, already shown; and from their never employing any party of respectability to manage for them in Assam; and having failed so far, shareholders refused to pay up any further sums on their shares. There was the most gross mismanagement. Servants of the lowest capacity, who, for a small salary, were willing to risk their lives in that sickly country, were alone sent there, to manage themselves and the Assamees, of whom I have given a description; and of the funds, $700,000 were spent in building steamers, &c., with the view of monopolizing inland or river freight from the Government, instead of being devoted to tea cultivation.

In the Tartar or Singphoo country, the only planter was Mr. Francis Bonynge, who had 4 plantations.

KAMOUN, NORTH WEST OF EAST INDIA.

The Hon. East India Company were the only planters.

Another wild scheme was started last year, in which the writer was asked to take part, but declined. It has since been placed under the management of Mr. Fortune, who visited China, and who has no further knowledge of tea matters.

This is the part of India where there is the least rains; and tea will not do without moisture. Planting in that district will be a failure. The Hon. East India Company have been at it for 15 years, and have done no good.

JAVA—PLANTERS.

Another wild scheme was set on foot by the Dutch Government of advancing money to planters (natives and Chinese), and taking the crop from them at 5 cents per lb. The illiberality of Government defeated itself. They were like the man who wearied himself to raise a plank, forgetting that he was sitting on it all the time. So far there is an account of tea cultivation in different countries. The only white man who ever cultivated tea out of China, on his own account, is the writer of these papers.

CAUSE OF RETIREMENT FROM PLANTING.

The Hon. East India Government induced me by letter, promising me a grant of Koojoo, Buramanjan and Gin-lang, and protection for myself and people, to enter the Tartar country—of a part of which they had taken possession. On the strength of these promises, I proceeded to the country, with the view of civilizing the people, and also to better myself. I worked hard in that out-of-the-way country (which although larger than some of the U. States, has not yet found a place on the Maps of the World), for five years. During that time, the Tartars took up arms to drive the British from the country, but proved unsuccessful. However, Government, for cause or causes not assigned, and without any notice to me, withdrew the guard from Koojoo, and also the surrounding guards, and so resigned the country, to all appearance; and the Tartars, who viewed me as the then sole representative of the Company, holding their land on the Company's authority, assembled at night and destroyed my property, and killed several of my servants. It may be seen by the following letter to Capt. David Reed, of the Bengal Artillery, that I had a very narrow escape.

The following is an article from the Calcutta Star :

"We had indulged a hope that the Singphoos were sufficiently convinced of our military strength at Ningrew and Rungagora Stockades, supported by the stations of the Dibroo Ghur and Jaipoor, to warrant the supposition of their never again resorting to their former predatory habits; but we are grieved to find another instance on record which proves we were over sanguine. Two of their most formidable chiefs, the Beesa Gam and Ningrew Lah, met with summary punishment, only last year, from the political authorities, on being detected in the perpetration of crime under aggravated circumstances, and promoting dissension amongst those tribes of the Singphoo favorable to British rule. Whether their period of confinement has terminated, or whether from lenient motives the political authorities have relieved them from that confinement, we have not heard; but it appears strange that a circumstance of the nature now alluded to, should have again transpired at a place semi-distant from the Stockades of Ningrew and Rungagora, about 15 miles each way—and, if we mistake not, an actual stockade itself, Koojoo—where, we know, some few months back, troops were stationed, as a piquet, close to the tea plantation of Mr. Bonynge. This gentleman, supported by a respectable agency house here, suffered similarly about the middle of his crops in 1844, fortunately for him without personal injury—in anticipation of an attack, perhaps, he having but just commenced a march into Rungagora when the Singphoos arrived. The whole of his property was destroyed, including a very elegant library; and we have been put in the possession of the manner in which everything was disposed of, as witnessed by a hidden spectator. Balls of opium, farm yard stock, vessels of different kinds, were forthwith despatched to their villages as booty of the first order. A cask of rum regaled the invaders (and wines and beers not being appreciated, the bottles were emptied), and the chairs were duly appropriated to the purposes for which they were intended. Contentment soon turned into mischief, such as cutting up books with their dahs (swords about 3_2 to 4 inches broad at the top, and tapering gradually to the handle, and about 26 inches long), and relieved tables and other furniture of their legs. Mischief quickly ripened into quarrels amongst themselves—some fighting, some stealing, and others committing every description of violence inside the burgalow (house), whilst those without were busy firing the whole. To a suit of clothes, everything had disappeared. On that occasion no produce was lost, Mr. Bonynge having cautiously, crop by crop, despatched it to a Government factory for final preparation; but this season we are led to believe that all has been sacrificed—400 monds (32,000 lbs.) Still, whatever the loss may be, we hold that after former warnings, and promises of good behavior to the Government, Mr. Bonynge will have a claim, inasmuch as he placed implicit faith on the mutual good understanding, and the tacit promise of protection both to life and property. Summary punishment must be inflicted on these races—by nature robbers and murderers—ere they can fully appreciate the British character. . Lenient and undecisive measures will not only encourage these attacks, but strengthen the belief, where it exists, of our arm being but a feeble one."

Last year, their villages were fired, their crops destroyed, and threats of utter annihilation proclaimed; those proximate to the nearest military station (Jaipoor) sought protection around it. Whole villages swarmed in, and settled on the Boore Dihing river, forming New Juggoon, Kujoo and Ningrew villages on unemployed lands; but the younger chiefs, indignant at the punishment of those more powerful, the Company Sahib Log (so the natives of India call the British East India Company), the desertion of their principal villages, and consequent want of laborers for tillage, will, in all probability, go on pursuing harassing midnight attacks, until a check is effectually put in force; and, for the safety of the traders, both Assamees or more enterprising foreigners, we would recommend, that when a possibility of a capture occurs, one and all of them should be removed, either to some spot on the Burampootre or elsewhere in our own possessions, as were the Gohains (princes) on the surrender of the Muttuk country.

The escape from death appears to have been miraculous; and it was fortunate for Mr. Bonynge that he commanded sufficient presence of mind to lay hold of the choppah bamboo (roof) while the Singphoos passed under him; equally fortunate was it that the absence of light precluded too minute a search; for, being a cripple at the time, and overpowered by numbers, he would have, had he been traced, undoubtedly fallen a victim.

We subjoin particulars of the attack from yesterday's "Hurkaru newspaper:"

* * * * But now for the escape which, without cant, I must credit Providence for. It was most miraculous. In the first place, I went to bed without loading my guns. A while after, a Singphoo teklah (pike bearer) commenced singing; he did the same the night of the robbery of the 529 rupees. I was so forcibly struck

with the supposition that something of the like nature was then exciting him, that I got up, called for a light, charged my gun with three pistol balls in each barrel, with shot to fill up the crevices, and drew the blind from over the door, so as to enable me to see out. I went to bed, fell asleep, and awoke about two o'clock. I turned round, looked to the door, and saw the Singphoos mounting the verandah (raised piazza) with a cat-like pace. I started out of bed, and cried out "Singphoos," when they cut down the teklahs (two men on watch, but who had gone to sleep). I then gave them a barrel; the room was small, and there were immediately about fifteen men cutting away at my bed, when I discharged the second barrel among them. I thought to reserve the second and work away with a dah, but, on laying hold of the blade, it was whipped out of my hand, giving me a slight scratch, the only one, except a slight spear wound in my knee, that I received; finding this loss, I was obliged to discharge the second barrel, when the whole gang rushed out of the room, leaving a man struggling on the bed, evidently in the last gasp. I did not lay my hands upon him, nor did I feel any relish to do so. I was about to charge my gun while alone, but on looking through the wall, saw so many Singphoos, and they actively engaged looking for a light, that I thought it best to escape; so I threw down my gun, and got over the mat wall, and clung to the roof until the Singphoos passed under me, when I got down beneath the "chang" (floor raised on posts) with the pigs. I knew that was no resting place for me, so I crept out on my hands and feet until I got the bathing-house between me and them, when I made the best use of my heels. I had been a cripple for the five days before; however, I had no time to think of bad legs, and ran well until overtaken with cramps. The feeling I had during the attack was, that the moment was to be my last, and I felt a savage determination to make the most of it. When a little way from the house I got a cramp and rolled over. On rising up again, I saw two men quite close. I knew I could not get away, so I drew myself up against the matted brush wood, and darted on the first, who fell in a lump under me. I had his dah (sword) half drawn ready to strike him, when the second came up and cried, "Chaprasee, sahib!" They turned out to be my own teklah and Captain Butler's chaprasee (chaprasee, a police court messenger, and teklah, a pike bearer), and my meeting them saved me from a jungle death; for the road or pathway (made by wild elephants through the underwood) was so full of water I never could have found my way to "Ningrew." It was as dark as it could be, and pouring rain in torrents. All is burned down; my pony is at Ningrew; he broke loose and would not allow the rascals to lay hands on him. The only thing I brought away was the night-shirt on my back.

I should say several Singphoos were wounded; but no person knows anything, all my servants having run away when they heard the gun.

One of the two men who I so fortunately met with, I started off to the nearest military post to call out the military, the other, I kept with me: from frequent attacks of the cramp, and my naked feet being full of thorns from the cane briars, and the painful state of my legs, I was obliged to sit down in the forest. I made the servant sit down, much against his will, with his back to mine, and tied him to me with his sheet, and placed his sword between my knees; in that position we fell asleep; awoke by day-break. The night was grand in the extreme, it poured rain in torrents; and every now and again the dazzling flash of lightning would pierce through the darkness of the forest; a tree would be upset,—its crash, and the breaking of branches of other trees, sent their echoes through all parts of the forest, setting the orangs and monkies to cry out, in the peculiar voice of the former, and the chattering of the latter; the trumpeting of the wild elephants, and again the piercing shriek or bark of the hog-deer, as she was startled by the noise, or surprised by the tiger, gave so much life to the wild desert, and the darkness of the night, that the fear of the pursuing Tartars, the recollection of the loss of everything I possessed, gave way to admiration of the war of the elements, and the confusion of the wild inhabitants of the woods,—and fatigue soon closed my eyes in rest.

The following morning a jemedar and his company, (a native officer with forty men,) started for Koojoo, but, as I could not keep on the pathway, I missed them. The Tartars were in pursuit, but being under the influence of rum and opium, kept talking, which gave me and my servant an opportunity to avoid them. We arrived safe, late next evening, at the military post of Ningrew. When I got in sight of it, through the reaction of the mind, the painful state of my legs, and the great loss of blood from leeches, my strength gave way, and the servant, by hoisting his sheet on a branch of a tree, signalized the guard, who sent out some of the soldiers to carry me in. Such was the unfortunate occurrence that ended my tea cultivation. Some of my people were murdered, and several wounded. See following extract from Dr. Sherlocks, assistant surgeon, East India Company.

"The Coolies (laborers,) who have come in wounded from Ningrew, are in great danger of their lives. The individual who has submitted to have his arm amputated, will probably recover; the other refuses to submit to an operation, and will certainly not live many days."

The following are letters of gentlemen, on that attack. From Col. Francis Jenkins, Gowhattee, Governor of the North East Provinces:

"The misfortune that befell you, it was not in the power of the local officers to prevent or remedy.

"I never ceased to regret the circumstances which caused you to retire from manufacturing, as it always appeared to me that of all persons who have embarked in those speculations, you were the most likely to succeed, and to do justice to our indigenous tea, and our capabilities of raising China teas, by your unremitting activity, and the ability you evidently had evinced. * *

"I can assure you I shall welcome your return to Assam cordially, for I know of no one more likely to turn our resources to account."

GOWHATTEE, January, 1849.

Major John Bracken, Assistant Adjutant General, wrote:

"I became acquainted with Mr. Bonynge in Assam, and whose unmerited sufferings while there, and the energy with which he bore up with them, engaged my pity and esteem."

Dr. William B. O'Shougnessy, Assay Master of the Calcutta Mint, wrote:

"I most strongly support everything Major Bracken writes. Mr. Bonynge is a most deserving man, and if you can help him, you could not aid one more worthy of it."

G. F. Edmonstone, Esq., Governor Cis Sutledge Territories, wrote, in complying with a request of mine:

"I do so the more readily, knowing you are an honorable man, and the victim of official displeasure."

Dr. George Lamb, Physician General of India, wrote:

"Government owes you something for your losses in the Singphoo country, where you were sacrificed."

The following letter is from a merchant of Liverpool, Mr. Daniel Willis, to Edmond Molyneux, Esq., British Consul, Savannah:—

Dear Sir,

The object of this letter is to present to you the bearer, Mr. Frank Bonynge, a young gentleman for whom I feel a peculiar interest, and in whose behalf I am proposing to solicit your friendly attention. My acquaintance with this gentleman has not been of long standing, and I therefore consider it right to give you a brief outline of his history, and the object of his present journey to America, in the pursuit of which he may, probably, have to ask your advice.

Mr. Bonynge went out to India some years ago, and after spending some time in the manufacturing of indigo, proceeded to Assam, under the recommendation of the East India Company's authorities in that quarter, to cultivate the tea plant. In this scheme he was engaged some years, with a prospect of considerable ultimate success, under, I believe, the advice of my brother (now a resident in Calcutta for upwards of thirty years), until a rebellion broke out among the natives, which was attended with the loss of all his property embarked, because the East India Company have refused to replace their authority over that district, or even to give Mr. Bonynge compensation.

Under these circumstances, Mr. Bonynge, with a view to introduce the cultivation of the tea plant into America, and under advice and letter from the American Minister, Mr. Abbott Lawrence, with whom he has been some time in correspondence, has undertaken to visit your Southern States; and, in the event of that failing to prove successful, I have suggested to Mr. Bonynge to employ his time in America in the acquirement of a full knowledge of the culture of cotton and tobacco, and their entire management. With the view of promoting this object, I have therefore to request you will kindly afford to him any advice which it may be in your power to give, should he request it; also, to supply him with letters to any parties who you think may be able to promote his objects, and disposed to advise him as to the best mode of doing so.

Any assistance which, in this way, you may render will be gratefully felt by the bearer, who, although unfortunate and ill-used in his career, is, I have every reason to believe, a most deserving person; and, in consideration, I need hardly assure you, that it will ever afford me sincere pleasure to reciprocate your good offices in this quarter.

I am, dear Sir, very truly yours,

DANIEL WILLIS.

KINDS OF TEA PLANTS IN DIFFERENT COUNTRIES.

All are aware that China alone supplies the world-wide with the little tea that wide world uses. But before entering into the nature of the China plant, it would be necessary to cast a look to those localities out of China where the tea plant may have, by some agency or other, made its way.

In Assam, the plant has been found indigenous; some say it has been carried there by those rivers running from China, such as the Boree Dihing; others suppose that man has introduced it, in his westward course. It is not very material to the purpose to inquire into the matter; but if it be of any interest to Americans to know how the tea plant extended from the west of China to the east of the Burampooter—an extent of territory that lies directly under our feet—I can do no more to help them to that knowledge than to offer suppositions. That the Singphoos must originally have come from Japan is but certain. There are another people called the Shans, and also another called the Fake-alls. The Shans and Fake-alls are the same, I believe; all came from the east, and all use tea in every possible manner. They put the leaf in their paun* and chew it; they boil the green leaf and drink the infusion; they beat the green leaf hard into a bamboo, and keep it there without any drying at all, and use it as required; they make extracts of the leaf, and keep it in leaves, or bamboos, or cups, in a thick consistency, in color and appearance like pitch. All easterns

* Paun is made of the betel leaf, with nutmeg, cloves and pepper, and lime made from shells.

5

drink their tea without milk or sugar. Whether these people carried the tea seeds from Japan into China at a remote period, or found it in China, and carried it with them in their progress westward, is hard to say. Gibbon mentions the Shans and their eruptions, in his history, to the westward. The appearance of the tea plants in the different localities west of China would impress the belief, that it has been always indigenous there, being found in the deepest forests, where man could not have resided for centuries past. Therefore, if it was man, man's profligacy has exterminated him, and Providence has preserved the plant. Again, if it was conveyed by water, and squirrels and birds helped to convey it miles from that water, and to places some three hundred feet above the reach of water, it must have taken an immense time, for large districts of country frequently divide one tea locality from another. Assam country is marked on the later maps of the world very irregularly. The Singphoo and Camptee countries, or Naga country, find no place on them. In the Singphoo and Naga countries the best tea is found. Down lower, i. e., to the westward, in the valley of Assam, the plant is inferior; the leaf is coarse, the tree stunted, and never reaches one-third the size it does in the abovenamed countries. The East India Company cultivated a baree (so called in Assam), or plantation, called "Chubwah," in the valley of Assam; the produce was good, being 320 lbs. the average per acre, but the leaf had an unpleasant, oily flavor about it. The company has given it up. All the tea plants along the Tingri river were inferior, and poor sterile plants; and the teas the Assam Company produced are of very little superiority to the present teas of commerce, which are far inferior to what they had been. My Singphoo teas were far, very far, superior to my Assam teas. In Assam I had four plantations. The first crops of leaves were passable, but the others felt like chaff in the hand, and would not of themselves fetch a very fair price. In the Singphoo and Naga countries it is quite different; the tea is of the highest possible flavor, requiring but one-half the usual quantity of leaves for infusion.

The British East India Company introduced China seeds; though they had greater facilities for doing so than any other government, yet the seeds they got from China were worse than valueless, for they all germinated, and grew up into a pretty little bush of two to three feet high, full of leaves, and no end to their powers of producing seeds. But the leaves were small, dry, hard, and so stiff that they could not be manufactured. From several thousands I had, I never made a pound of tea. The Assam Company tried every scheme; even plucked off all the leaves, supposing, when the new ones came out, they would be soft, but they came out so slowly, and still so dry and poor, that the experiment proved a failure.

In Kamoun mountains, north-west of East India, the British East India Government established a garden in 1835, under Dr. Jamieson. Up to 1850, they have exported no teas. They made some from some Assam plants, and sold them to the natives of the place. They have frequently endeavored to sell their plantations, but no one would buy them. These are the plantations Dr. Royle lauds so much.

There is another party recently entered into cultivation in the same district, and who offered, or rather a friend, Mr. Prideaux, offered for him, to place the concern under me. I declined, feeling that there could be no success. Since then, I see from the newspapers, that Mr. Fortune has been to China for seeds, and brought round some 1,700 in a budding state. I wish him success of a dry climate and China seeds.

The Washington Government has commissioned their men-of-war to bring home tea seeds from China; hence the necessity of this article on the qualities of the tea plant. Government will, of course, distribute the seeds to all parties; they will sow them; they may, or may not, germinate. If they do not, so much the better; because further expense and labor will be saved. If they do germinate and grow up, years and years will roll on, and the little tree will grow up to three, possibly to four, feet in height, bloom, and produce seed. The seed will be carefully saved and sown, and so the country may become extensively provided with tea bushes, and will not be one pound of tea the better. As far as I can learn of the cultivation now being carried on by Mr. Junius Smith, LL.D., from his own letter, and other people's accounts, he has unfortunately fell into the evil against which I would caution all. For Mr. Smith's part in tea affairs, I am not surprised at the little advance he has made. Unacquainted with the cultivation, picking up his knowledge from magazines; and by compilations from them making up a pamphlet, he reasoned himself into the belief that all was fair and square before him; and with a spirit of enthusiasm and enterprise, he resigned his long-loved smokes of London, and the thin and highly-polished pumps of the citizen, for the great thick brogans of a clodhopper, and transported himself into the wilds of Grenville, there to ruralize and plant tea. Such ardor was highly deserving of success. His prospect, I fear, is the poorest. Mr. Smith is now going on four years in America; and from his letter may be collected his failure. His letter is dated 4th July, '51, directed to the editors of the *Journal of Commerce*, and in which he wrote:—

"On that day, (4th July,) I plucked from several of my green tea plants, a small quantity of tea leaves. *The small number of my plants, and the partial growth of the leaves*, forbade my attempting to gather beyond a sufficiency for experiment, but enough I apprehend to confirm and establish the important fact, that the tea plant of China is congenial to our climate; that the tea is pure American growth, unmixed with any herb or material; that it is cured by solar heat alone, and is in every respect the genuine tea of China tea plants. *Its fragrance, flavor, and physical qualities, may undoubtedly be changed by the process of manipulation and manufacture.*" There are two things I would advise Mr. Smith to do, viz: If his mode of manipulation and manufacture has undoubtedly changed the fragrance, flavor, and physical qualities of his teas, to give up manipulating and manufacturing, for if the physical qualities and fragrance and flavor be changed, what other property of tea remains to be changed, I am at a loss to know. And secondly, if after the time he has spent he finds his tea plants few in number, and *his tea leaves of a partial growth only*, to get rid of his plants also.

Java.—It was impossible for the cultivation of the tea plant to succeed under the unfavorable patronage of royalty; no planter, except on a most extensive scale, could sell his teas for 5 cents per lb. I have never been able to ascertain if other causes operated against its success in that Dutch possession.

DIFFICULTY OF GETTING GOOD SEEDS FROM CHINA.

I WILL now add the evidence of others, who it will be seen have had long experience in China, to show the difficulty of getting good tea seeds from that country. One of them, Rev. Mr. Gutzlaff, is known all over the civilized world by his works, and the other, Mr. Ball, was a tea merchant and tea taster, to the honorable East India Company, and for twenty years a resident at Canton, and whose word is of all others the most worthy of confidence.

Mr. Ball states that "It has been observed that the Chinese universally agree from remote antiquity to the present, that only the Bohea mountains produce the highest flavored teas; they moreover affirm that it is only in the central division of these mountains, which are known to the Chinese by the appellation of Vuy-Shan, (inner mountains,) where the highest flavored teas are produced, and that the tea deteriorates in quality, till in some of the remote districts, the leaves are thin and poor, and of no fragrance or sweetness in infusion; that no labor can make them good,"—and, that "The Ankoy teas, grown in the vicinity of Amoy, are for the most part inferior; and the Honan and Waping teas of Quang-Tong, (Canton,) may be given as examples of still greater inferiority. And again, says the same author, "the vast inferiority of the flavor of Twankay tea, the product of the green tea hills, with little or no attention paid to its culture, to that of the Hyson tea, the product of the highly cultivated plains, would be apparent, and interesting to any one disposed to try the experiment at the average prices." Also, Ball still goes on to say, "The towns which extend seventy ly (twenty-three miles) from Vuy-Shang, are called Py-Keeng, Tien-Cza-Zy, Tang-Moa-Kown, Nan-Nang, &c., the leaves are thin and small, and of no substance, or whether green or black, or made with much care, yet have no fragrance. Tea is also produced as far as Yen-Ping, Shan-u, Gen-Nong, Kien-Yang, Hew-Shang, and other places, but unfit for use." I might still go on further with Ball, but I feel the reader will be satisfied. Gutzlaff says "The extent of the soil that produces the best Bohea, is no more than forty ly," but this statement is wholly incorrect. And in another place adds: "In all other parts of Fo Keen where the tea plant is in a similar soil, and under the same climate, it never thrives so well as there (Bohea Hill.)"

It is seen by the foregoing that there is no good seed to be had at Ankoy, (Amoy) or at Canton, nor at Shanghaie even; that the seeds the Chinese would give, even if they had good ones at hand, would be the most inferior; but the good seeds are not at hand; and, as there is no means of knowing the seed of an inferior plant from the best, money would not procure the better ones. But the jealousy of China is too well known, to suppose she would part to foreigners the means of shutting up one of her two greatest exports. It would be altogether sanguine to expect it. Her export of tea is about 26,000,000 to 30,000,000 of dollars, and is it to be supposed that a nation like her would throw away that export? It must be remembered by all who have read the history of the silk worm, that it was conveyed out of China in a cane-stick by a Jesuit.

There is no admittance for strangers into China ; a few years ago, half a dozen Englishmen went up the river from Canton in a boat, and went on shore. They were all murdered. Mr. Fortune states in his book that he proceeded as far inland as Soochoo, but Mr. Martin states that he failed in the attempt ; and from his description of tea making, and his statement that the tea plant must be renewed every four or five years, there seems to be internal evidence in his own work, that he has been laboring under some hallucination on the subject.

As he has now proceeded to Kamoun to plant tea plants, where he will find plants now in bearing, some of them fourteen or fifteen years old, and which will bear for several years to come, he will find out his error.

GEOGRAPHICAL EXTENT OF CULTIVATION OF TEA IN CHINA.

Valleys, and table lands, and round the bases, and partially up the sides of hills or mountains, seem to be favorable sites for tea plantation. To ascertain what is going on in the interior of China is very difficult, the Jesuits and other Catholic Missionaries alone obtained ingress into China—and except their attention be drawn to the particular information required, they deem such matters out of the scope of their calling. Therefore, it is only by running over the various authors who have written on the subject that information can be obtained, and then it requires an intimate knowledge with the tea plant itself, to be able to sift the probabilities from the improbabilities; these travelling authors write at such random, as if their sole and only view was to make up a book, and not to benefit mankind.

The tea plant is raised on the Island of Chuzan for domestic use, the price they could obtain from the China tea dealers being so illiberal as not to pay them for their labor, which, probably, they may better employ in fishing, &c. Mr. Bell, who visited Pekin with the English Embassy, found the tea plant, kept there as a curiosity, and about, to use his words, "as big as a currant bush." Pekin is in N. latitude 40°. In N. latitude 29° the bush has to be covered with straw, and tied up with ropes, to protect it from the snow and frost; well, all that labor would make the tea so much more costly. And as the mountains are continuous northwards, it is natural to suppose the cold of winter increases, and that the tea plant could not be at all productive for commercial purposes. I will give here the following extracts to show the face of China, taken from McCullough's Dictionary of Geography:

"The mountains and hilly districts of China comprise about one-half of its area. A portion of the Great Mountain system of East Asia entering this country, on the N. W. and S. W. frontiers, subsides previously to its termination to the coast into low hills; so that, tracing their course backwards from east to west, they gradually ascend in terraces or slopes, and give to the south and west districts a mountainous, and to the east division a hilly character. Northwest, at about 34° N. latitude, and 102° E. longitude, the great Pe-Ling range, which has already traversed a portion of Thibet from W. to E., is joined by the Yun-Ling chain, which, entering China at about 31° N. latitude, and 101° E. longitude, descends southwards nearly to the province of Yun-Nan. These mountains, from the easternmost edge of the high table lands of E. Asia, are snow-capped, and inaccessible to the natives, being actually left blank on the Chinese maps."—DAVIS.

Another ridge joining the Pe-Ling at the same point, takes an opposite or N. N. east direction, and entering the Empire in the Province of Shen-Se, reaches nearly to 110° E. long. Another arm of the Pe-Ling—the Ta-Paling chain—intersects the country from west to east, to about 115° E. long.; the Pe-Ling itself continuing in its former course, gives out various branches which traverse the central provinces. The other mountain chains join the stupendous Himalaya ridges, and enter into the country at its S. west in the Province of Yun-Nan, from whose high table lands the most extensive Chinese range rises. The Yun-Lung, the most southerly of these chains, runs nearly east into the Province of Quang-Tong. But by far the most important mountain chain is the Nang-Ling, which branches off from the northern edge of the Yun-Nang highland, runs eastwards to within 150 miles of Canton; it then inclines N. E. to its termination to the harbor of Ning-Po, having given out many branches, some of the mountains belonging to which, rise above the snow line.—MACARTNEY EMBASSY. Most of the mountains here mentioned end in low hills in the Eastern Provinces, which consequently comprise the hilly districts.

It will be seen from the above extracts, that one-half of the area of China is covered over with mountains.

That the S. west portion is cut off by the great Pe-Ling, which enters China at 34° N. lat., and the Yun-Ling entering in at 31° N. lat., and descending southward to the Province of Yun Nan; and another ridge in the opposite direction, N. N. east, entering Shen-Se. The Ta-Paling intersects the country from W. to east to 115° East long., and the Nang-Ling runs off from the north to 150 miles of Canton, and then inclines N. east to the 30° N. lat., or Nan-Po, having given out many branches, some of the mountains belonging to which rise above the snow line.

The above will show, that if in 29° of N. lat., or green tea district, so much care has to be taken of the plant, that upwards, or at least north of 30° N. lat., owing to the great altitude as well as the increasing cold from the natural causes, that the country is unfavorably severe—and hence the plant becomes a green house curiosity at Pekin, 40° N. lat. Again: If we proceed southward, we find that the Nang-Ling starts off from Ningpo 29° 30′ N. lat. and long., 121° 30′ until it nears Canton by 150 miles. Canton is N. lat. 23°, and long. 113° 25′, so that we are cut off by that diagonal range reaching above the snow line. If we go down to Amoy, N. lat. 24°, Ball says the teas, for the most part, are inferior. But the price current will show them to be almost valueless. And again, the Waping and Honan teas of Quang-Tong, in which is situated Canton, 23° N. lat., "may be given as examples of still greater inferiority." Below, or south of the 27° N. lat. there is no good tea produced; and, in fact, N. lat 28° and 29° and part of 30°, may be put down as the good tea districts. Gutzlaff says, "the extent of the soil, the best Bohea, is not more than 40 ly,

(13 miles) in circumference," but that must be a random statement of the reverend gentleman, because, if we take his other statement to be true, which I believe, and have from the Chinese themselves, that no plantation exceeds a few acres. Therefore, 13 miles in circumference, or say $4\frac{3}{7}$ miles in diameter. of that, or about 16 square miles or 10,240 square acres—at the rate of 300 lbs. an acre, would only produce 3,072,000 lbs. of good Bohea for the consumption of all China and outside Barbarians—whereas, her exports alone is thirty times that, and I should say the exports is but a small fraction of her own consumption—as the Chinese will pay $7\frac{1}{2}$ dollars for the best tea. However, there is no good tea produced below 27° N. lat., nor is there any exported from north of 30° N. lat. And we see these latitudes are traversed and retraversed by immense mountains, as they extend to the west to Thibet, across about 18° of longitude.

It is well known, too, that no tea is produced near the coast in these latitudes, and that all teas have to be carried to Shanghaie, about 260 miles. And furthermore, it may be pointed out, that there are many districts in these four degrees of latitude, where the tea grown is of no fragrance or flavor.

"The Chinese say," writes Ball, "that the tea deteriorates in quality, as the plantations diverge from that particular (Bohea Hills) locality, till in some of the most remote districts the leaves are thin and poor, and have no fragrance or sweetness in infusion, that no labor can make them good."

The teas of the neighborhood of Canton, (Honan and Waping), Mr. T. A. Gibbs, in his evidence before the Parliamentary Committee, states, is such, that no process of manufacture would render it suitable for the British market.

The Thea Virides, (a fancy name of botanists), says Mr. Fortune, in writing from Chusan, is cultivated everywhere. . . Every small farmer or cottager has a few plants on his premises, which he rears with considerable care; . . and although the shrub grows pretty well, it is far from luxuriant, as it is in the large districts on the main land. Chusan is N. L. 30°.

"Every province," says Ball, "probably produces much of its own teas for common domestic purposes, though not for festive or ceremonial purposes." . . . "The Chinese say the provinces of Py-chy-ly, Shang-shy, Shew-sy, Shang-tong, and even Honan, are unfavorable for tea; and in the official returns, there is no mention made of any plantations in Keang-nan."

So that it may be seen, whether the exports of tea be East from Shanghaie, or South from Canton, or N. West by caravans, that the tea is supplied from 28° 29° and 30° of N. lat. Thibet ought to be, and I have no doubt is, a more favorable country for tea cultivation than China.

Adulteration of Tea.

It will be almost invariably found, in taking tea leaves out of a tea pot, that there can rarely be found one that has not the appearance of the tea leaf. I have seen a few exceptions, where the tea had an unpleasant flavor of tar, or something of that kind, about it, in which there were other leaves than that of the camelia kind.

However, there are so many leaves closely resembling those of tea, that there is a very wide scope for adulteration, without resorting to the clumsy expedient of mixing leaves of every tree or bush that might be at hand, and so render detection an easy matter.

"The teas of the districts of Py-Kien, Cza-Sy, &c., &c., &c., the leaves are thin and small, and of no substance, and whether green or black, or made with great care, yet have no fragrance. This, however, is used for congous. Tea is also produced as far as Yen-Ping, Shang-u, &c., &c., &c., and other places, but is unfit for use; there is reason to believe that tea from these places are constantly mixed with low congou, and that many of the congous technically termed "faint," come from these places, as will be seen by the accounts received from other Chinese, where some of the above places are enumerated as producing tea forming a part of the tea exported as congous."—*Ball.*

Mr. Fairbridge states in his evidence before a committee of the British Parliament, "that the better kinds of teas have degenerated." And Sir J. Francis Davis, the British Minister at Canton, says, "two kinds of bohea are brought into this country (England) from China. The lowest of these is manufactured on the spot, therefore called Canton Bohea; being a mixture of refuse congou with a coarse kind called waping, the growth of the provinces."

(I call most respectfully the attention of the U. S. Government to these pages, and beg them to reflect on the serious injury they will inflict on the country by the dissemination of China tea seeds, such as the Chinese will give). "Again," states Sir J. F. Davis, "the consumption of bohea in this country (England) has of late years increased to the diminution of congou, and the standard of the latter has been generally lowered."

And again Sir J. F. Davis stated, "the young hyson, until it was spoiled by the large demand of the Americans, was a genuine delicate young leaf, called in the original language Yien-Tsien, before the rains, because gathered in the early spring. As it could not be fairly produced in any large quantities, the call for it on the part of the Americans was answered by cutting up and sifting other green tea through sieves of a certain size."

All the old leaves are of a reddish color, and require some coloring matter to make them green or black. All the black teas are classed as follows:

1st	quality, Flowery	Pikoe; 2d of that quality is Pekoe.
2	"	Pouchang.
3	"	Souchang.
4	"	Compoi.

5	"	Congou.
6	"	Bohea.

These may have different names, as Yen-Pouchang, &c., from places, or from the times of the season they are made in; for both soil and season will affect their qualities; and would, therefore, render it necessary for Pouchang or Souchang, &c., to have different significations, to denote such differences. The bohea is a very old leaf, and the substance in a manner is dried before the leaf is collected from the tree; and, in fact, what is now called congous, is little better. These reddish leaves are colored black; sometimes with the sap of a particular tree, and sometimes by green grass burned, and various other ways. The dealers in London may, for all I know, use Day & Martin's shoe blacking, for their black teas is to be seen in the windows, of a polished black. All that would of course give the tea a deep color when poured out; so it suits the eye for sale, and it suits the eye in drinking.* I do wonder, however, that doctors often forbid the use of tea, although they must know that the unadulterated is anti-bilious, and the most healthy beverage that can be made use of. Milk and sugar help to change its qualities. It is only white people who use either. In fact the Chinese and people of the N. East of India consider milk an exerement, and will not use it.

* "I visited a 'tea manufactory' a few miles north of Canton, where about 500 men, women and children were engaged in converting coarse looking refuse leaves into several sorts of green tea." —*Martin.*

PRICES OBTAINED BY THE CHINESE PLANTER.

I HAVE stated in the foregoing pages, that between the planter and the purchaser of teas for America, there were a great many who derived their bread by selling it from one to another, who may be called the middle men of the trade. The people who make their living on tea-trading in China are innumerable. And through the agency of whom the price of tea is raised to 20 cents by the time it reaches Canton, and 16 cents at Shanghaie, per lb. average.*

If it was possible for an "outside barbarian" to calculate, and deduct the profits per lb. and expense of inland carriage, to either Canton or Shanghaie, it would be in reality found that the cultivator was very poorly paid.

First, there is the rent of land, calculated by Gutzlaff a $1 50 per acre. There is the interest he has to pay on advances to him (to purchase corn, &c., which his land under tea might have produced him,) 75 per cent.† Then say a poor man has an acre of tea, and that it produced him 300 lbs., and that he realized 7 cents per lb., he would only have 21 dollars; that would be a poor means of support—especially when the interest of money, government revenue, and rent of land were paid off.

Now the Hong laborers, or Coolies, get 15 shillings sterling, or 45 dollars yearly, and their diet and lodgings.

The shoe makers, according to Rev. Mr. Smith's evidence, get nearly the same wages.

The tea packers, are allowed about 36 dollars a year, and diet.

* Average price of the teas exported to England.

† The legal interest is 36 per cent.

So the Chinese planter would not be paid proportionably to that value of labor.

The expense of carrying tea from the tea district, is 5 to 6 cents per lb. Then comes government duty, 10 per cent., say 2 cents per lb., and say five or six parties who turned it over from one to the other, made in the aggregate 8 cents per lb., for if any one will observe the fragile make of a tea chest, he will understand that in passing over hills, and loading, and unloading, and exposure, that there must be a great deal damaged, and that not less than probably 10 cents would cover the risks, and give so many parties a little profit. Therefore, there is duty per lb., 2 cents; transit expenses per lb., 5 cents; to parties trading, say 8 cents; or 15 cents per lb., leaving the cultivator to get only 5 per cent.

Mr. Fortune says of the tea planters—" They may be considered in the same light as our vintners, poor, hard-working people, gaining just so much as is requisite for their daily subsistence. The cottages amongst the hills are simple in their construction, and remind one of what we used to see in Scotland in former years, when the cow and the pig lived and fed in the same house with the peasant."

Gutzlaff writes, " It is to be remarked that only the poorest lands are allotted to this cultivation, the rich being allotted to the growth of rice and sweet potatoes. The answer of the people, when questioned on the subject, has always been, that the profits were more uncertain, and generally less, and therefore it was much better to make a sure livelihood."

Therefore to two parties injustice is done; to the planter and the consumer. This leaf which can be produced from 2 to 4 cents per lb., by the time it gets to Canton is 20 cents; and by the time the consumer gets it, it stands him on an average, 100 cents per lb. adulterated, old, and compared with the China tea leaf, used in China or Russia, valueless. Such is the imposition on both parties by the number of middle men who live on the planter and consumer.

However, I believe things are brought to that pass by the ingenuity of traders during the last two centuries, that we can get none of the good teas,—truly and really the tea generally drank at Calcutta, England, and America, is not good, nor has hardly I may say any relationship to good tea—that the good teas are drank by the Chinese—that they are sent west to Thibet, Nepaul, &c., and Russia. Ball says the Yen-Pouchong cost from 15 to 30 shillings, or $3 75 to $7 50 per lb., and the price of tea in the Bohea district, which is classed as Pouchong and Souchong, is from 2s. 1d. to 4s. 7d., or 52 cents to $1 15 per lb.

The China cultivator on such teas may be remunerated, and then the people that are imposed on, are the English and American consumers, who have to pay 100 cents per lb. for Chinese refuse teas, and tea house sweepings, mixed up with worthless leaves.

PRESENT EXTENT OF THE USE OF TEA, AND THE PROBABLE CONSUMPTION IN TIME TO COME.

If we consider China, and find in that sole tea producing country 367,000,000 of inhabitants; if we consider these as tea drinkers, and contemplate a Chinaman and his family in easy circumstances, all about him particularly clean, and tastefully arranged, with his cheerful round face and small eyes; looking to one side of the apartment, there may be seen a little tea pot, and beside it a tea cup, to which the whole family have recourse at meals (three times a day), in common; and in the intermediate time, as the one or the other may be thirsty; or as a neighbor visits, the pot is again in requisition. Such tea drinkers cannot use, at the lowest calculation, less than 20 lbs. of tea a year each; but 30 lbs. would be nearer the mark.

Now the Chinese who has not tea will buy tea; the woman will card, spin, and weave her cotton into cloth, that she may increase its value, go and sell it, and buy tea for the price, or oftener barter it for tea. Then may be asked where ends the number of lbs. consumed by the above number of people?

Probably the quantity now exported by sea to all places may be under 120,000,000lbs. I should say the mark would be 96,000,000 to 100,000,000lbs; that by caravan to the N. West of China it is impossible to know. It is not easy for a white man to pass through Nepaul; the Nepaulees and the Gourkees are daring hill people, and wild. The Thibetians are like the Chinese; and to pass through the Singphoo, Mesmees, and Camptees, would be impossible. So to ascertain what the export in that direction is at the present day, or to form an approximation to what it may be, is out of the question; Martin puts it down at 10,000,000 lbs. If it be said the Chinese are tea makers, and that consequently they must consume a great deal, that is not a sequiter. It may be recollected of Grimaldi the clown making a tour to Newcastle to eat fresh salmon: when he got there, to his disappointment, he could get none, as they had all been preserved and sent off for sale. This may be the case in some measure with the Chinese—some may sell their goods, but all may not do so. If it be said a Chinese would consume 20 lbs. of tea in the year, then the inference would be 7,340,000,000 lbs. If it be said some may not use it, but as a great many make it their daily beverage, probably it might be reckoned that an average of once a day for all would be a fair calculation—well that would be still 2,527,000,000 lbs. Now, according to Rev. Mr. Gutzlaff's statement, the extent of the district for producing good tea is only 40 ly, or 13 miles in circumference, which would at the most only give 3,072,000 lbs., leaving an opportunity for any other nation to supply China with from 2,527,000,000 to 2,700,000,000 lbs. of good tea. This will show two things: the difficulty to say what quantity is consumed in China, and the random statements writers are capable of making. It is impossible to say what quantity of tea is consumed in China, Nepaul, Thibet, &c., &c.

All that is known for certain is, that by twenty-two millions of Americans there are 20,000,000 lbs. of tea consumed; that in Great Britain and Ireland, by 30,000,000, there are 50,000,000 lbs. consumed.

Extent of the use of tea under a new state of things is to the purpose to inquire; and I will recapitulate the expense of producing tea to that end.

In Assam—
Plantations, Selling and Bally-jahn, per lb. $3\frac{1}{2}$ cents.
In Singphoo country—
Plantations, Koojoo, Bura-man-jan, Gin-Long, . . . per lb. 2 1/[illegible] cents.

Probable expense in China,	-	per lb. 4 to 5	cents.
Java contract price for delivery,	-	per lb. 5	cents.
Probable expense in America,	-	per lb. $2\frac{2}{3}$ to $3\frac{3}{5}$	cents.

The first question then to consider is, is tea necessary to man? To a very large portion of the human race it is a necessary of life. It may be replied that a man can do without tea. So he could; he could do without coffee, without meat, without fish, without milk or butter or cheese, yet all these things are necessaries. To the Chinese, habit has rendered tea necessary. Tobacco is necessary to the habitual smoker, and see to what lengths he will go in the use of it.* Tea is necessary to every tea drinker; and perhaps there is no period in the twenty-four hours when a man, or the family of a house, are so sociable and cheerful as at tea time. Genuine tea is a healthy beverage, corrective of a tendency to over secretion of bile; and there is in it a very soothing quality, which refreshes both mind and body, without bringing on after depression of spirits. These qualities may be in some way interfered with by the large admixture of other leaves, by coloring matter, or by milk, which is so liable to grow sour.

If the poor Chinese woman did not think tea a necessary for her, she would not card, spin, and weave cloth and sell

* It is a well established fact, although not probably well known, that tobacco contains one of the most powerful poisons—nicotine—in quantities from 2 to 9 per cent., five drops of which would kill even a dog. From a great deal of experience, I can say that a tobacco smoker has but a poor chance against jungle fever. Out of all who I know got that fever, none recovered but myself and two others. Two of us did not smoke; the other smoked very moderately.

it for tea; nor would the poor laboring classes in this and in European countries, who work so unsparingly of themselves, pay 75 cents to $1 for a pound for it.

Well, it is seen that China cannot produce enough of tea for herself, and that the poor in Europe and America, the Indian, &c., cannot get an article, that cost only some 2 to 5 cents to produce, for less than on an average 100 cents per lb. for an adulterated kind. Therefore, tea is in a manner forbidden to man, by man's own mental indolence, and by the reprehensible indifference of those who have had opportunity, and means, to investigate all things connected with it. Truly, man counteracts the benevolent designs of the Creator, and then blasphemes against Him, and accuses Him of the miseries of which his indolence, his apathy, and his selfishness alone are the causes.

Let the consumption of tea in England be taken into consideration—30,000,000 of people consume 50,000,000 lbs. of tea, for which they pay from 100 to 150 cents per lb. The duty levied on tea by the English government is 55 cents per lb., although some of that tea sells in London as low as 8 to 12 cents per lb., or one-seventh the amount of the duty alone—so the poor hard working man has for that tea which sold in London for 4 to 6 pence the lb., (in bond,) to pay the extraordinary price of 3s. 6d. to 4s. per lb., or (87½ to 100 cents per lb.) Can anything be more oppressive and tyrnanical, can anything be more cruel, than that a poor simple laborer and his poor family should be so shamefully mulct—that poor man whose sinews are ever on the stretch, whose perspiration is ever pouring forth for the support of the whole fabric of society—for the support of that wretch who rolls in his carriage, and spends his life in idleness, corrupting and debauching all who he comes across by his bad example; and is a curse and a stigma on his species?

Of England and Ireland's 30,000,000 of inhabitants, say—

500,000	drink tea twice a-day, or 15 lbs. yr. each,	7,500,000 lbs.
4,000,000	" once " 7½ "	30,000,000 "
12,000,000	" once a week—1 lb. per annum,	12,000,000 "
	Total, - - - - - -	49,000,000 lbs.

This shows that only 4,500,000 enjoy the daily use of tea, that the other 25,000,000 can have no tea except in case of sickness or on festivities. I may say, and would be near the fact, if I stated that 4,000,000 of tea drinkers alone consumed the 50,000,000 lbs. There are very few tea drinkers in England or Ireland, who do not take tea morning and evening. Coffee is used more in coffee houses than tea; the coffee is 1d. to 1½ pence per cup; tea, 2 pence to 2½ pence per cup, both bad as bad can be. The coffee in many cases is made from crisped peas, toasted bread, even of things disgusting to mention. Tea is equally adulterated, or substitutes used equally dangerous to health, in all the London coffee houses. All that chicanery and fraud in the tea and coffee trade, in tea and coffee houses, which goes to injure the health and constitutions of so many citizens, could at any time be easily remedied in the last one hundred years, can be easily remedied to day, if people will inquire, and be not ignorant.

A man who drinks tea once a-day, consumes 7½ lbs. yearly. Twice a-day, 15 lbs. yearly—

Say, 15 lbs. of tea at 100 cents per lb., - -	$15
To every 1 lb. of tea 8 lbs. of sugar would be used, or 120 lbs. sugar, say 10 cents per lb., besides milk, - - - - - - - - - -	12
To one person only	$27

a year for tea and sugar for a poor man, who has his tea twice a-day. Will the friends of temperance look to this, and consider the expense to a poor family—not only in America, where there is more employment and better wages, but in the mother countries, where labor is at a low price, and the poor man's means scanty to a painful degree.

It can be seen from the above, that the great bulk of the people is denied the use of tea. And therefore it is that some 100,000,000 lbs. are only consumed in all Europe, America, and all the English colonies.

Supposing one-half of the population of China drinks tea once a day, i.e. 180,000,000, at 7½ lbs. each—1,350,000,000 lbs. Then take all the east, Thibet, Nepaul, Burmah, Siam, British East India, Persia, Turkey, all Europe, Russia, America, including north and south, Africa, Australia. In all these countries, more or less tea is drank. There are, alone, 200,000,000 of people in British East India, all of whom would drink tea, if they could obtain it. To say, what may be the extent to which tea may be consumed, would be impossible. But to suppose, that if the people of England and Ireland could have genuine tea at 20 cents per lb. instead of the mixed quality they now get for 100 cents, it would not be unreasonable to suppose, under the circumstance of tea being so placed within the reach of the whole population, that then there would be some 150,000,000 lbs. consumed.* The French are the most economical people in the world, they will not now use tea. The only tea imported is calculated at 300,000 lbs. Reduce the price of tea and the case would be different. And in America, U. States, say 22,000,000 consume only 20,000,000 lbs., hardly that. Let her produce her own tea, and the case will be different. There is no article of consumption for which there is so extensive opportunity for progressive increase, and that in the eastern world too, more so than in Europe and America. Europe, Asia, and America contain to-day 900,000,000 of human souls, nearly. If 150,000,000 lbs. are consumed by thirty millions of people, it would not be too much to say, that the whole population above would consume 500,000,000 lbs., and that America would find ready sale for that amount at 20 cents per lb., which would be equal yearly to 100,000,000 dollars, or twice the amount of the whole cotton crop in value. But if it be calculated that England, when she can get her teas at one-fourth the present cost, that then she consume 100,000,000 lbs., only twice her present quantity, which would be a very low calculation of 3⅓† lbs. a head—then, of the 900,000,000 of population mentioned, it would only require one in every six to drink tea, to consume 500,000,000 lbs. However, America herself, at no distant period, will consume 500,000,000 lbs. of tea, if she will but set to and grow it, and from the universal desire there is for tea, it

* The Americans consume upwards of 140,000,000 lbs. of coffee.

† Much less than the people of Jersey and Guernsey consume, who consume 4 lbs. 4 oz., and Australia 7 lbs. per head.

will be the one article most extensively consumed. America can, if she will. She has the soil, climate, energy, and intelligence to appropriate to herself the trade.

STATE OF TRADE BETWEEN CHINA AND ENGLAND, AND AMERICA AND CHINA.

Mr. Ball has shown the expense of China teas when they reached Canton and Shanghaie is per pecul (133⅓ lbs.)—20 teals, 2 mace, 3 cash, and 17 teals, 7 mace, 4 cash, or about 1s.0⅛d., and 11¼d., or 25 cents at Canton, and 22½ cents at Shanghaie per lb. Now, the average sale price for 1849 was only 20 cents at Canton, and 16 cents at Shanghaie. This will show, if Mr. Ball be correct in his statements, that a very heavy loss must be sustained, and is being sustained, on the China side of tea matters. However, adulteration may make up the deficiency.

But, for a series of years, trade in China has been unfortunate for the English. The Parliamentary Committee of 1847 declared that the loss to England, taking the trade both ways, was 35 per cent.; and in the returns of the trade with the five ports of China for 1847-48, laid before Parliament in 1849, by order of her Majesty, Mr. Macgregor states:—"From the information which I have gathered, I am led to consider that all the shipments of tea during 1847 has had a very unfavorable result for those concerned in them. This is fully borne out by the fact, that while prices at home have been progressively declining, the prices here, particularly of the common and middling kinds of congou (which form the great bulk of export), have been maintained at the same point to which they were when these teas sold at 25 to 30 cents higher in England." In what manner the trade may have been rectified in 1849 and 1850, would be difficult to tell. Prices in England have been much the same.

Of course, China has the ball at her foot. She will not go out of her own port to buy or to sell. She sees England, the United States of America, France, the East India Company, &c., &c., waiting at her door for her to buy what she may require, or to sell what she pleases. England is trading with her at a loss of 35 per cent.; possibly America, too, may be obliging to her. However, America has been more careful than England; the latter finds it difficult to slacken her pace in any trade, to the injury of herself and others. Both nations are forcing their goods and their money on China. England sells her $10,000,000 worth of goods, and gives her $10,000,000 of cash, for tea and silk, a little sugar-candy, and camphor. America gives her all her goods, with some 6,000,000 dollars besides in cash, for tea, silk, &c. Then, in steps the British East India Company, laughs in his sleeve at the English, American, and Chinese, slily points at 100,000 chests of smuggled opium behind the scene; and says, "Well done America! well done England! Contribute your shares to China, that she may add them to her own means, and buy these 100,000 chests, and so supply us with an addition to our Indian revenue of £5,000,000, or $25,000,000, upon which our political existence depends." The English merchant cries out—"Let us buy more tea from China, and she will buy more of our cotton piece goods." The American says—"Cannot we increase our trade with China." And the Chinese cry—"Come, John Bull! come

Yankee! Ye outside barbarians, buy our teas and our silks. We are out of opium; no smoke since morning!" And both parties hand over in cash 16,000,000 dollars, and nearly as much more in goods; and then sweep across the mighty waters with a collection of *forest leaves; and that is the way to increase the trade with China.*

It was argued, on the part of England, if we can induce the Chinese only to wear a cotton night-cap each, 367,000,000 night-caps! Why, the whole of Lancashire would become one continuous city. Manchester and Liverpool would kiss each other; and all England would have been under one perpetual cloud of factory smoke. And what a day that would be for cotton in America! Ambitious men would pitch all idea of a future presidentship of the South Atlantic States to Old Nick, and scamper off to pick cotton bolls! But the disobliging Chinese will buy opium, which he, in his humble ideas, considered more inducive to a nap than a cotton night-cap. More than that, they have the audacity to go almost naked, or wear a sheepskin turned inside out, and even to grow cotton, spin, and weave it, and sell it to buy opium, and even tea, from his neighbor. Yes; let America push on her cotton at an undervalue of 30 per cent. to what it had been some years ago, and let England push on cotton piece goods made from that, at a loss of 35 per cent., upon a trade, taken both ways, all to cap the Chinese and clothe the Indian. Yes; let England and America be the servants of servants; and, as they are the boasted first-class of the human race, show these Mongolian races that they will forget the bases of all permanent wealth—cultivation—to dress and deck them out with fancy robes, at such losses, forsooth, to have a great trade!

The trade with China has been long forced. It is seen that both nations give her upwards of 30,000,000 dollars, more than one-half of which is in cash; that she only takes some 13,000,000 dollars from them of their wares; and yet both England and America say, "Come, let us buy more silk, more tea, and she will buy more of our cotton goods in return." Although both parties see they already favor her yearly with some 16,000,000 to 17,000,000 of dollars in cash for her teas and silks, and that she hands over that with some 36,000,000 to 38,000,000 of dollars for opium, they still will persist in giving more.

CULTIVATION OF TEA IN AMERICA WOULD BENEFIT ENGLAND.

England cries out, let us reduce our duty on teas, then we will consume more of China tea, and China will be able to buy more of our "piece goods." Reduce the duty 6 pence or 12½ cents per lb., that will increase the consumption in England fully 10,000,000 lbs. See Parliamentary Report of '47 on commercial relations with China. That would be enriching China! Did ever any person hear of 367,000,000 of people being enriched by 30,000,000? Yea, by that 30,000,000 consuming an additional 10,000,000 lbs. of tea, which at 20 cents per lb., would be 2,000,000 dollars, or about $\frac{5}{9}$ of a cent per head, for Chinese to buy piece goods with.

However, were England to reduce her duty to-morrow, from 55 cents to 5 cents per lb., what would be the effect? Say that England then consumed 200,000,000 lbs. of tea, 150,000,000 lbs. of an increase, it would be 150,000,000 lbs. more than China could supply. The present exports of China are say 100,000,000 lbs., and for a great portion of that she is indebted to spurious leaves; for every pound she had to supply last, or this year, she would have $2\frac{1}{2}$ lbs. to supply next year. Would the Chinese, who is in the habit of drinking tea from his childhood, give up the use of it for 20 cents per lb., which does not, it seems, pay the Chinese. The Chinese who, as has been seen, pay for Pouchong and Souchong kinds in their own country, from 52 cents to 115 cents per lb., and for Yen Pouchong, from $3 75 to $7 50 per lb., give up the use of it for 20 cents!!!

Would the smoker give up his tobacco for the usual price? Of course not. And if the Chinese were asked to sell his teas for his own consumption, he would naturally ask upon what grounds? And if he saw an urgent demand for tea, he would be urgent in holding out for higher prices, so that, as the demand was increased, the price would go up. That is the natural sequiter in all commercial matters—and instead of the consumer in England deriving the benefit of the reduction, it would go to the Chinese.

Tea is not like other crops. It takes three years before any material quantity of tea can be had of tea-trees, and then, if a sufficiency of trees and lands be planted to produce a quantity to meet the demand, from the small quantity obtained the third year, the increase of the third, fourth, and fifth years would glut the market; well, if to avoid this the Chinese only planted moderately, so that the increase on the fourth year would be sufficient to meet the increased demand, then for that four years the revenue would go to the Chinese; but it would go for a longer period, for the Chinese would not commence planting until high prices induced them to do so; that would take a year, and then they might look upon the rise as temporary—and considering the necessity of their land to them, it would be very difficult to make them throw it out of one cultivation, which paid them yearly, into another that comparatively would not pay them for three or four years afterwards. Therefore it may be reasonably supposed that tea would be high in price for five to six years after reduction—probably for eight to ten years.

The revenue the British government derives from tea, is £5,600,000, or $27,000,000. The import of tea is generally 55 to 57,000,000 lbs.; if only 50,000,000 of that be consumed in England, and the remainder re-shipped, the duty on 50,000,000 lbs. would equal $28,343,750—and to reduce their duty from 55 to 5 cents per lb., would be to lose in a manner, if not in revenue, at least to revenue and consumers, the whole of the ten parts out of eleven of the above twenty-eight millions of dollars. China will not increase her plantations without a cause. Plantations raised, and from which tea is not manufactured yearly, would be but a waste of land, as the trees would have to be cut down to make them productive when required; and the plant from the seed would be nearly equal to it in point of time, and in point of endurance far superior. Then England stands in that peculiar position; she must go on as she is, until some country cultivates the tea plant, and be able to furnish her with the teas she may require in addition, or sacrifice for years her revenue of some 27 millions of dollars. And in her embarrassment, she would, even if she had a disrelish so to do, be obliged to take American teas, if presented;

not so, however, she would feel it a pleasure, to be able to get her teas at one half the price for the consumer, and in largely reducing her duties, still by the increased supply and consumption, keep up her revenue to the present point—so that England would in reality get twice the quantity of tea she gets now for the same money, or for less even, and maintain her present amount of revenue.

Indigo.

On this article little need be said. It has been cultivated and manufactured in the Southern States heretofore, and is now grown wild over them. But unfortunately, there is nothing but cotton, cotton; every blessing and every advantage that might be reaped, is ingulfed in the only one and all-engrossing idea—cotton. When indigo was an export staple from these States, the trade was very different from what it is now; and it is extraordinary, and shows the indolence of the human mind, that in the Southern States there is not one to be found who has ever inquired how the indigo plant cultivation has been getting on since it was carried away from America; what revolutions the trade in the article has undergone; what the consumption now is, and what its consumption was in the time America was the sole producer of it; to what amount it is manufactured now; and what prices it realizes; what the advantages may be that the eastern world may derive from its cultivation. No, there is not one that has even dreamed of being inquisitive on the subject. Indigo in East India realized some 15,000,000 of dollars. The amount of sales of all cotton produced yearly the last twelve years, would be 53,000,000 dollars; yet here is an article one-third the value of that upon which the whole of the United States are dependent for their prosperity; an article all know can be cultivated here successfully; and yet there is not one to ask the reason why it is not now cultivated. Yet hardly a newspaper can be taken up, but there is a puny scheme advanced for the relief of the cotton planter. The gossip all are content with is, "Our fathers cultivated the indigo eighty years ago. The British Government, to whom we were then colonies, gave us a shilling a pound bounty for producing it. Our fathers could make out but a poor pittance on it, and, therefore, it was abandoned for cotton, for which we got one dollar to one dollar fifty cents a pound." Well, the day is gone by when cotton would realize that price, and now the planter would be but too glad to realize one-half former prices. Therefore a great, very great change has come over the days of the cotton planter; it was but this very year when it was the merest chance saved all concerned in cotton from a general bankruptcy.

On the other hand, what has been the course of indigo cultivation and trade? The export from the United States some 60 years ago was some 134,000 lbs. only, and sold for a price of 2s.6d., or 62 cents per lb. There is now an export of indigo from East India of 13,000,000 lbs. which is sold on the spot (Calcutta) for a price, the lowest years, of 100 to 140 cents per lb., and in the highest years 200 cents per lb., and the lowest description 60 cents per lb. Some of it has sold as high as $2 45 per lb. Such are the changes that have taken place in the two articles; one, cotton, has sunk to the lowest rate it can safely be produced for—the other has risen to a price which gives to the planter in a single year a fortune.

It will be very probably stated that the United States could not produce indigo as cheap as East India, which would be as much as to say, whenever East India chooses she may take our cotton cultivation from us also; we cannot produce it as cheap as she can. Cotton planters will not wish to acknowledge that; but, while it is true that they can, and it is to be greatly feared will, take away the cultivation of cotton, yet, from local cause, America can take away from India the indigo trade.

Let the reader take the map of India, or of Bengal, and look upon its face. It will be perceived that the country is one continuous flat, covered over with paddy districts and indigo tracts, dotted all over with indigo factories. It will be seen the whole of that country appears like an anatomical drawing of the human arteries and veins; and it will appear a wonder how it is possible to travel it, from its being so cut up and intersected with rivers. Look at the great Ganges flowing down from the North West; look at the mighty Burampooter sweeping down from the N. N. East; look at the Soan River, and see them all joining their ocean of waters, and then sub-dividing into thousands of streams, called the mouths of the Ganges. Travelling there is in boats, of which there are various descriptions, the Bugerow, the Boleah, the Panswah, &c., &c. If the traveller be in one of these boats at any time from May to the end of October, in his wanderings he may, in the evening, see the country high and dry before him, and villages reposing in peace and security. In the morning he may be riding in his boat where one of the villages lay, and behold on the spot its wreck, the dead bodies of its inmates and of its cattle floating around him. This is not unfrequently the case. Wet days, or the snows thawing on the hills from an unusual hot day, fill up the beds of the Ganges or Burampooter, which collect the rains of the surrounding countries, carry them down on Bengal, and cover over the face of the country. Should these floods commence early, the fate of the indigo planter is sealed. The prospect he had on going to bed of realizing a support for life, in the morning is converted into the prospect of a residence in debtor's prison. Or he may get up in the morning and see all lost by evening; or he may watch anxiously the rising for three or four days; his plant is not ripe to cut, nor can he obtain boats even so, to convey it to his factory.

Besides, there may be no rains for the October sowing, nor none to save the April sowing—all then is lost.

Beyond these local reverses, which make the planter consider himself a most fortunate man if he can get two successive profitable seasons to pay off his heavy liabilities, and give him the means of retiring; or even fortunate if he can get one good crop in three years to keep himself afloat; there are other causes, which make the production of indigo in the East comparatively very expensive. The planter, although so called, is not a planter, he is a manufacturer. The natives are the planters—and the only inducement to the natives to plant indigo for the Englishman is, he gets money in advance for the indigo plant which he is to deliver months afterwards. That advance enables the native to accommodate himself in other things,—such as sowing paddy or planting sugar-cane, on both of which he makes a profit; on the indigo plant he makes little or none, and, as seen, the whole may be a loss, as the poor ryot of India is so destitute that he is unable to pay the planter. Therefore he sows the indigo to discharge himself of the obligations incurred by taking advances, and not from the remuneration from the plant itself. Consequently many of those who take advances never will sow; and the majority of them will not sow except by compulsion of the law. Every advance made has to be given on government stamps, or the government will, if at all, assist the planter with great reluctance. But beyond all these troubles it often happens that the planter will resort to every expedient to compel the riot to take advances and cultivate the plant; and for that purpose he rents the districts around him from the zemendars, (land-holders,) that he may have the immediate control over the ryot, (farmer,) as his landlord, and through that power force him into the cultivation; and to obtain that power he will pay the government revenue for the land, and pay another revenue to the zemendar. The trouble ends not here; the lands for indigo are limited. The zemendar often sells his right to two planters; but generally all zemendaries (estates) are held in India by a family, each having a certain share, not in the division of the lands, but in the proceeds. One brother may have one "hissa," (share,) another five, and perhaps some friends or cousins five shares more. Each may go and sell his right to different planters, who make their advances, and when the crop comes to be cut, then is the "tug of war." Lattimars (hired fighters) are called into requisition, broken heads and bones, and frequently murders, are the consequences. Then law-suits on law-suits, appeals after appeals; and there is but little exaggeration in stating that of the whole cost of the production of indigo, 25 per cent. goes in law; and certainly 20 per cent. more in advances that never can be recovered. And it is usual on reading an advertisement of an indigo factory for sale, say for one which may be valued at 30,000 dollars, to see the statement, the balances due on the factory of money advanced, of 40,000 or 50,000 dollars. The value of indigo factories is fictitious. The buildings of a factory worth 30,000 dollars might be worth possibly 400 dollars (not including the residence, which is sometimes a palace, sometimes a bungalow, which may cost for building 300 to 500 dollars). Indigo presses are nothing more than a few strong posts of iron with screws on the ends, and nuts, with a wrench to screw down the nut, and may cost 100 to 200 dollars, for a set; the best indigo boiler would cost 100 dollars, or perhaps 150 dollars. Therefore the capital required to establish a good large factory where a planter could make some 200 monds (mond of indigo called factory mond is 75 lbs., the bazar mond is 80 lbs. The bazar mond is the one generally used except in the case of indigo,) which would sell, if good quality indigo, for 15,000 dollars, at present prices, if very good or best quality, nearly 20,000 dollars. The whole of that factory could be constructed in America, buildings, presses, boilers, &c., for 1000 dollars, or less.

I cannot give the produce per acre of indigo. From our mode of conducting business in East India, it could not be obtained without some difficulty. The ryots cultivate sometimes in little patches, sometimes they join in cultivating some extent of land, the measurement of which they know not, and is only ascertained on another occasion, that of paying rent for it. The measurement is per " biggah," which varies very much ; and to ascertain the produce per biggah, they do not know the quantity of land at the time of disposing of the produce, and probably men who are unacquainted with accounts do not remember the produce when they are made aware of the quantity of land by the call for the rent. They leave part of the plant also standing for seed.

The plant is sold in Bengal by the bundle, which is measured by a chain. In the Doab it is sold for 1 rupee, (50 cents), for 5 to 6 monds. 200 to 225 monds of plant to a mond (75 lbs.) of indigo, is a fair average produce. Therefore it would cost about 36 to 40 rupees, or 18 to 20 dollars, for the plant necessary to make 75 lbs. of indigo. The expense of manufacturing would be but little.

The water in India for indigo steeping is drawn up from wells or rivers, either by the China wheel, or by bullocks. The wells are some of them 70 feet to 90 feet deep, some 20 feet. Throughout India the water is not more than 14 or 16 feet from the surface, except in unusually high altitudes.

A vat would contain 90 to 110 monds of plant. To supply three vats, it would take in India, from a well, two bullocks and a man a day. In India, the wages of the man and his bullocks would be 8 annas, or 25 cents; two men to fill and empty the vats with the plant, 6 pice each, or 3 annas, equal about 10 cents; six men would beat two vats, or say nine men to the three vats, at 6 pice each, 13½ annas, or say 40 cents; one man to boil six vats at 12 pice, or three annas a day, one-half equal to 5 cents; one man to procure wood, 3 pice, or 3 cents; two men to press it, 3 annas, or 10 cents; packing, one man to 3½ monds would be but very little, say one-third of 5 cents. Conveying the plant from the field would depend upon the distance; the indigo field is sometimes four and five miles from the factory in India.

I will give the above items in tabular order, with an estimate of the probable expense in America:

Item	Cost	Total
Cost of 200 to 250 monds or 16,00 to 20,000 lbs. plant, say - - -		$36 00 to $40 00
Three men to fill and empty 3 vats	15 cents.	
Raising water for 3 vats -	25 "	
Half of one man's salary to boil -	6 "	
Nine men to beat 3 vats -	45 "	
Two men to press the indigo -	10 "	
Expense of conveying 200 monds, say $2	00 "	
Fireman - - - - -	5 "	
Wood - - -	30 "	
Packing and chest, 60 cent. 3½ monds	20 "	
	$3 56	$3 56 to $3 56
Total expenses per 75 lbs.		$39 56 to $43 56

To which is to be added expenses of law-suits, loss of advances—making it at the very lowest 53 dollars.

PROBABLE EXPENSE IN AMERICA.

It is necessary to ascertain in some way the produce per acre. Thirty monds would be a good produce per biggah; the biggah measures 20 khudams (steps) of five feet each; the step in India, or khudam, is the space between where the right foot is raised from the ground, to where it rests on the ground again—twenty khudams, equal, therefore, 100 feet; that squared is 10,000 feet—43,560 square feet in an acre—therefore 4⅓ or more biggahs in an acre, and consequently there would be 130 monds, or 10,400 lbs. of green plant on an acre. The biggah was generally calculated five to an acre. The Bengal biggah is three to an acre.

But as the above is my own experience in measuring and weighing, I will here follow it. Now the ground where I had been cultivating that indigo was excessively sandy—so that at the lowest calculation 130 monds, or 10,400 lbs. of plant, may be put down for an acre in America.

For indigo I would give five men to prepare an acre and sow it, not that the labor is greater than in cotton; weeding, one man; cutting the plant, six men per acre; the conveying it to the factory would cost little, as the factory could have the lands around it under indigo, which could not be the case in East India. Therefore,

For preparing and sowing land,	6 men per acre,	at 20 cts.	$1	20
For weeding	2 "	"		40
Cutting plant	6 "	"	1	20
Conveying to factory, a man and horse, say - -				60
Two men to fill and empty one vat - -				40
Two men to beat two vats - - - - -				40
One man to boil six vats, $\frac{1}{3}$ part of his wages for two vats				8
Firewood, and man, two vats - - - - -				28
Packing and chest, $3\frac{1}{2}$ monds, say 60 cents — $\frac{2}{3}$ -				20
Raising water, two men for six vats—for one vat -				7
			$4	83

As 220 monds of plant make 75 lbs. of indigo, therefore as 130 : $4 83 : : 220 : or $8 17 per mond.

This is not much more than one sixth the price it would cost in India. In America, all the beating of vats and raising of water could be done by machinery. The sowing of indigo would be from 1st of April, and the manufacturing would end the middle of September. The indigo plant requires to be only weeded once, and there can be no hoeing after the seed be sown. If it is shown that the manufacturing with labor at twenty cents in America, is cheaper than in India where labor is put down at five cents, it arises from the purchase of the plant. The indigo fails so often in India from causes shown, that if the riot did not get a fair profit when successful in saving his crop, to pay for former losses, he could not go on.

Paying for labor fifty cents per day, the expense of 75 lbs. would be

Preparing and sowing land,	6	men per acre	$3 00
" weeding	2	"	1 00
Cutting plant	6	"	3 00
Conveying to factory			1 00
Vats, filling and emptying,	2	"	1 00
Beating vats	2	"	1 00
Boiling			10
Firewood, &c.			25
Packing and chest, 3½ monds, 75 cents, ⅕			15
Raising water			20
			$10 70

Say 220 monds to 75 lbs. of indigo. Therefore as 130 : : 10 . 70 : : 220 : $18 10 for 75 lbs.

The lowest description of indigo sells in Calcutta for not less than 30 dollars for the 75 lbs. The average price for good, for the last years, would be about 65 dollars for 75 lbs.; but the best Bengal indigo is rarely under 80 dollars, and from that up to 100 dollars. Some time ago it had been up as high as 340 Rs. or 170 dollars ; that is, the sale price obtained by the planter at Calcutta, for 75 lbs.

It will be seen the advantages possessed in America, from the latitude of New-York down to the lowest point in Florida ; for as the sowing and manufacture take place between 1st April to middle of September, throughout that space it is fully warm. The cultivation of indigo extends from Madras up to Delhi. Upwards, from Patna, the plant has to be constantly irrigated ; the hot winds that set in more or less in March, and April, and May, are indeed very hot, and there is no rain until 10th or 15th June. Irrigation is carried on from a well of some 12 to 16 feet deep, lined with great thick straw ropes towards the bottom. The well generally contains from 12 to 20 inches of water, which is drawn up in an earthen pot. The work is painfully slow. The expense of manufacturing, that is, the expense of purchasing the plant and manufacturing, is about 30 dollars per mond, which is considered very moderate. An European would have no success in cultivating the plant himself. He could not possibly expose himself in the hot winds.

America has her lands waste, and can build her factories in the centre of her fields ; and having done so, it will be found that my estimate is above the mark at least 20 per cent.

Preparing the land and sowing, weeding once, cutting the plant, filling the vats and taking it out again, is all the labor ; anything else is the work of one man to a large quantity.

PROCESS OF CULTIVATION AND MANUFACTURE.

The land is ploughed or hoed, say some nine inches deep, and the soil is pulverized, i. e., clods well broken, roots of grass and weeds carefully taken away; then the seed, mixed like flax-seed with clay, is cast in the ground, and a very light harrow; a bush with moderate weight on it is used often in India. If weeds spring up with the plant, it would be necessary to take them out; the plant after a few showers covers over the land, and keeps down all weeds. It grows even to some six feet high, varying from four feet to five feet. When it gets, or before it gets, to its full height, and before the leaves get yellow in the least, the plant should be cut, and carried to the factory the same day. All plants should be cut very early in the morning, and then placed in the vats, or otherwise not be heaped up to get heated. Each vat may be made to hold from 5,600 to 8,000 lbs. of plants. The plant is all placed horizontally in the vat, and when filled up, hurdles are laid on the top of the plant, and beams are laid across the hurdles; the ends of the beams being secured at the side walls of the vat. The water is then poured in, and the plant is steeped for ten hours or upwards, depending on the heat very much. The water is then drawn off from a vent at the bottom of the vat, into another vat, built at the base of the one in which the plant had been steeped. The beams are then raised off the hurdles, and the hurdles taken away; and the steeped plant is taken out of the vat and made use of for firewood. A large quantity of potash might be obtained from it.

The water being drawn off from the upper vat, the steeped plant is then beaten up by six men entering into it, and beating with their hands until the coloring matter which is contained in it begins to show itself in small atoms. The men then get out, and the indigo or fecula subsides, and soon after the water is drawn off. There are two vents in the lower vat; the upper vent is for drawing off the water, the lower one for drawing off the indigo, and a quantity of the water which could not be well drained off, without disturbing the fecula. The fecula is then put into a small vat, either of wood or masonry, and allowed to rest some time, and then more of the water is drained off. It is then taken to be boiled in a boiler generally from six to ten feet square, and four or five deep, and all froth carefully skimmed off. It takes five or six hours to boil it. The boiler is made of copper or iron, as the party may fancy.

When boiled, it is let out from a vent in the bottom of the boiler into a vat, where the fecula soon subsides, and more of the water is then drawn off. It is then filled into square cases, pierced with small gimlet holes at about two inches apart; in the wooden square is placed a cloth fitting to the square; and then the boiled indigo, still retaining a good deal of water, and consequently of a thin consistency, is filled into the square; a lid is then placed on the top of the square, which fits into it, and all is placed under the press; and as the lid is pressed down into the square, it forces the water through the cloth, and through the holes in the side of the frame; then, when all the moisture that can be pressed out is done so, the sides of the square or box are taken off, and the indigo left on what had been the bottom. The whole is then divided by a board, or measure, into eight parts, and cut through by a piece of wire, giving sixty-four squares; then each square or cake is placed on a hurdle in the shade to dry. The doors of the drying house are locked up, and the indigo in that state takes a month to dry; when it is packed in a strong coarse case, and sent to market.

In precipitating the indigo, it is not good to use anything. Lime is destructive, and gum makes it hard, and liable to crack, which is not liked.

Date Tree.—*Phœnix Sylvestris.*

It requires little more than to bring this tree to the notice of Americans. The cultivation of it is simple; any man who ever planted a tree can manage this. The tree gives toddy, (a milky kind of juice, and intoxicating,) on the third or fourth year. In India the planter taps it so soon, that the jackalls drink the taree or toddy out of the earthern pots.

This tree grows all over East India, but it is said to do best on the sea coast. It produces 8 to 10 lbs. of sugar, and even 12 lbs., per tree.

It may be planted one in every 12 feet square.

The sugar from the date tree makes a beautiful grain; it has, if exposed much, a peculiar flavor, and consequently in East India it is frequently mixed with cane sugar, which improves both.

The tree will grow in all the pine lands, and in all sandy lands.

To collect the taree, a slit is made in the tree, and a pot, (earthern,) is secured immediately under the incision, into which the milk is poured, and from time to time, as the man, or woman, or boy, goes round, the milk may be collected and brought into the factory. The tree gives no trouble beyond the planting, and, if it is required, crops of vegetables, &c. can be grown between the trees. It is one of the hardiest trees in India, and grows without cultivation. Wherever a squirrel, crow, &c. may drop a seed, there will grow up a tree. Since 1840 this tree has been very extensively cultivated in India, and a large portion of the East Indian sugar is made from it. There may be 300 trees or more planted to the acre. It yields no fruit, and may be looked upon as an ornamental gigantic sugar-cane.

THE COFFEE PLANT.

To introduce this plant successfully, would be doing a great service to America. The question some would put is, would it succeed in America? Many experimentalists would answer in the negative. However, there is an old saying, give the old fellow his due, and to a useful article, give it more than its due; give every care and attention to those productions that are so valuable. The consumption in America is 145,000,000 lbs. and in England is, say nearly 37,000,000 lbs., (this was the consumption in 1846); there can be no great increase in the amount, there being in the last years so extensive an adulteration of this article. France consumes say 50,000,000 lbs.* These numbers will show the quantity of coffee that must be used throughout the world, if some eighty millions of French, English, and Americans consume

* In 1841, the consumption was 45,000,000 lbs.

232,000,000 lbs., and some 70,000,000 lbs. of tea besides. A little plant, of so vital importance, requires consideration. Has it found any? What experiments have been tried? Have these experiments been published? Has any premium been offered, and how often, for the successful growth of the plant? I fear agricultural societies in America are nothing more than periodical gatherings of crowds. I have met with no agricultural society in America more than a few people who collected to discuss things that have been cultivated for scores of years, and old wines.

A man in America gets a little plant stuck in a pot, the pot is watered when it is convenient to do so, or when it is not forgotten. However, moisten the earth ever so often, is it certain that a few handfuls of clay cut away from its mother earth, (and from all those fluids and gases &c., and the changes the heat of the sun, or the dews or evaporations of the night may cause,) retain the same qualities to support vegetation? The plant derives its nutriment from the earth. There can be no doubt that there is a continued circulation of fluids and gases through the earth—and that a portion of clay or mould being removed and placed in pots, &c., is cut away from that circulation, and consequently the plant grown in pots has not the same means of nutriment that it would have in the earth—and therefore must be defective in its nature, and differ from the plant in the ground of the same species. Every man who takes upon himself to make these experiments should duly consider the value of the plant in hand, and should be most careful how he treated it—and should in such case, if one, or two, or half a dozen experiments were not satisfactory, still persevere, and try all soils in the locality, and plant seeds from all climates, and procure plants at all ages, from not one only, but from a variety of climates. There are cases where some plants cannot be raised from seeds, and where even young plants from other countries cannot live without changing their nature; yet old plants being introduced may succeed, and the seeds from such old plants after a few years may be acclimated, and fit for the propagation of young plants. However, I do not think that all that labor is necessary for the coffee plant. There is every variety of soil and climate in the United States of America, and abundance of room can easily be found adapted to the growth of coffee or anything else; and abundance of coffee seeds or plants can be had acclimated for at least all the States down from 35° of North Latitude. However, I do not recommend that any party should invest a fortune in coffee plantations; a large quantity might be produced without great expense by having a few garden trees or hedge trees, or half an acre or so. There are coffee plantations high up on the Chera hills of North East of India, where it is excessively cold in winter; they belong to a gentleman with whom the writer is personally acquainted. I have seen coffee grown wild in the 27th deg. North Latitude of India, on hills, some three to four hundred feet high, where it was intensely cold, and where there were frost, snow and hail, and all around it the higher mountains capped with perpetual snow. There is nothing therefore required but attention to the various sides of the question and perseverance. And it is not only coffee that requires these virtues, but every thing else—and without them there can be little success. It would be well for society if it could in some way be saved from the present inundation of books of light reading. It is not the time only that is spent in poring over them—but they render the mind incapable of devoting itself to the quiet, solid information that would be useful, and render every man possessing such knowledge a useful member of society. It may be safely supposed that one per cent. of those capable of reading, and who do read, do not give five per cent. of their leisure time to that useful study that would tend to promote society. A Mrs. Trollope—or any other old Mrs. that might write some exciting stuff—the divorced—the disappointed—the courtship—the newly-married—the rake—the suicide, &c.—will command more attention than all the philanthropists in whose hearts a holy love is ever burning, and whose heads are ever engaged in deep study for the good of their fellow-creatures. Were Father Mathew to write a book to-day, after all his heavenly ministrations, a book that would, if read, lead thousands to happiness—it is doubtful, if it was handed to any of our great periodical publishers, with the manuscript of some novel from the pen of any of the well-known novelists, whether Father Mathew's book would not be rejected, the novel published, and prove a good speculation. And this is the great improved society of 1851.

However, to return to my subject. The coffee plant should be introduced into America—not as a speculation by which fortunes are to be realized; any that would move in the matter for immediate profits, probably would be disappointed; but it is easy for any one possessed of a garden, to have a few plants, and in that manner let the growth proceed gradually in these States. I have seen a great deal of speculation in coffee planting, and invariably money lost. However, mismanagement had a great deal to do in the matter. The cultivation in Ceylon is beginning to pay now, and probably, with justice, will pay well.

I should say there is much injury done by experimentalists introducing at once exotics from very hot climates to colder; for instance, because it is near, from West Indies to America, to North of 32° of N. Lat. It is easily to perceive the great change of climate there must be in carrying a plant so many degrees of lat. from its native place, whereas degrees of longitude change the climate little; perhaps any change in such case would be owing to local causes. Every degree of latitude crossed, is a certain degree of change, except intercepted by great altitudes, &c.

Judicious proceedings and perseverance will introduce coffee. How has indigo been introduced, but by a young lady possessed of these qualities, Miss Lucas, who, like many young ladies determined to accomplish their wishes, prevailed over every obstacle? The first attempt of Miss Lucas was a failure, so was the second—the third was successful. I wish there were a few young or old gentlemen, or young ladies to-day, of Miss Lucas's determination and active mind.

Look at the cotton spinners of England in their endeavors to produce cotton in East India. They feel that all the elements are there for its successful cultivation; they will not admit of the idea of a failure. The British East India Company have expended some 100,000 dollars; it has got some half dozen planters from Georgia; it has told the people of England, "We have done so much to send you better cotton." The people of England say, "Yes, and you must do more." And the House of Parliament is moved, session after session, and the British East India Company charged with blame and reproach, and urged to proceed. Nor do the people stop there; they make efforts on their own part, and have sent out their own commission of inquiry, which will cost several thousand dollars; and if that should fail, they will send another.

That is the kind of spirit to make progress. There is intelligence declaring itself resolved to compel every obstacle to yield before it. That is the spirit that made a poor little island the most powerful people of the times; and it is the spirit that will force the productive powers of India, in a few years, into competition with the world. That was the spirit that raised up Carthage, Rome, Venice, &c., &c.; that was the spirit that, in modern times, raised up America to it present grandeur and greatness. Is that prosperity now to ebb back for want of that spirit? Americans, look to the falling off in your staples to the disappointment of all, as too certain indications of stagnation, and of want of that spirit in you.

THE MANGO TREE

Is the most important fruit tree that exists, and if all do not agree that its fruit is the very best, must agree that it is one of the very best. The fruit of the best kind is as large as the largest citron, but somewhat flat and oblong in shape, and measures seven to eight inches in length, and nine to ten inches in circumference.

It has a kernel inside which, in a fruit of the above measurement, would be about four inches long and an inch in thickness; the skin is about one-tenth of an inch in thickness.

The fruit is used in three ways—as mango fool, as pickles, and as a fruit when ripe.

The tree is a beautiful tree, with very thick foliage, acting as an impervious umbrella against rain, and a most delightful retreat from the heat; for in the heat a resinous substance is given out, which gives the sweet perfume of the fruit itself. It grows to about fifty feet in height, bears the fourth year, and as it gets larger in the sixth and seventh year, bears from 200 to 400 fruit. This tree is generally planted in India in groves, generally of twenty to fifty trees. The kind called the Bombay mango is the best. Up towards Delhi the best fruit is to be found; but in Dacca, Calcutta, Madras, and Bombay, the fruit is excellent. The mangoes are not all of the kind I have described. There are some nearly as inferior as those to be seen at Charleston from the West Indies. The West Indian mango has a nasty turpentine flavor, and is so stringy, that it takes a day to pick the fibres from between the teeth. There are none so bad in India, but there are some very inferior.

The mango fruit is an article of great inland commercial importance. In the great cities of India, for four months in the year, whole streets are lined with stores of them, and thousands of people are to be found hawking them about. Hackeries (carts, drawn by two bullocks) loaded with them, pour in from all parts, and the ghaut (wharfs) are thronged with boats bearing them to the markets. At the meat bazaars, to which all resort in the early morning, there is no provision basket leaves without its hundred or half-hundred mangoes. The poor live on them, and the rich indulge. Few there are who do not eat them some time in the day, most all three times—at breakfast, lunch, and dinner; and ladies retire to their apartments, that nobody may be the wiser how many they may eat.

If in the Mafuzil country a man should sally forth in careless wandering, he may see over that interminable plain Midan clumps of trees here and there to break the sameness of the place. He may wander from the scorching heat into one of their shades; his senses are in every way delighted; he is under the deep foliage of the mango tree; its perfume regales him, and all in and around is as if it were all creation associating in one united family. There lies the shepherd in the shade, his goats and sheep resting by his side. Even the wide-spread herd of antelopes come to sniff the sweet odor, and would fain lay themselves down were it not for man's presence. Further off skulks the savage wolf and his few companions. Around, from leaf to leaf and from flower to flower, the butterfly flutters and the bees display their varied hues, and hum forth their song. The grasshopper and the mole-cricket are found in every variety. There the crows sit picking the sheep; there the familiar mina hops from the sheep to the goat, and from the goat to the shepherd; and then mounts the crow's back, as if desirous to amuse his neighbors; there is the widow-bird floating through the air, with its long and graceful train; there is the brilliant jay, the golden oriole, the scarlet tuddy bird; there are the parrakeets; there is the sweet bulbul; there is the tree-duck and the tree-teal, and the smaller and larger green pigeons, and the timid hare gambolling at the other end. Such is the collection to be found in a mango tope. And then, there hangs the mango in all its golden richness among the thick, glossy, dark green leaves. And there, too, may arise a more holy remembrance, viz., that that grove had been planted by order of the sick old man on his death-bed for the improvement of his country, and in compliance with the desire of his Shaster (Hindoo bible); and who will say that that good act which gave shelter to so great and varied a number of God's creatures, and fruit to man, did not bring, on that feeble old man, God's mercies.

For America, the mango trees would be the greatest acquisition. They would be the best possible trees to line the streets of its young progeny of Babylonian cities. They would be a tree in the private gardens to which the owner would point with pride, and watch with the greatest vigilance, and send a few fruit to his long-respected neighbor with the greatest complacency. For the farmer and the fruiterer, the tree would be a source of great profit, and every family could add new luxuries to their table—mango fool, mango pickles, and mangoes themselves in season.

But poor indeed must the mango still appear from my description of it, to what it really is; and I submit that Americans ought to make some exertions to procure it, to see, to taste, and to pronounce on its qualities.

THE LEECHEE TREE, OR LITCHEE,

Is a shady and large tree, some 40 feet high, ornamental, and bears the fruit of that name. It is a delicious fruit, as large as a good sized plum. It produces a very large quantity of fruit, and there is not the least injury to be feared from a free use of it. In that respect, it is like the mango. The fruit is dried in India. Foo-chew Fo, in Fo-Kein, in China, is noted for her leechees, and her trade in them in a dried state is extensive.

The leechee tree is not of equal importance as the mango, but, as stated, it is of a delicious flavor, and plentiful in the season, June, July, and August.

THE JACK FRUIT TREE

Is also a large tree, from 40 to 50 feet high. The fruit grows from the stem of the tree. It is very large, some equal to the largest sized water melon. The skin is rough. The natives are very partial to it; and it is frequently served on the tables of Europeans in India, in pies, &c. The tree bears from 30 to 60 fruit, and as no care or attention is bestowed on the tree, and bearing so great a quantity of vegetable food, it is one of great importance with the natives. Another advantage is, the fruit arrives at maturity at irregular periods, i. e. one jack may be ripe, when another is but of a very small size on the same tree. I never was partial to the fruit; but once in my wandering through the immense and gloomy forest of the Singphoo country, with one follower, we found a jack tree, with one solitary fruit on it, which saved us from starvation, and enabled us to make our way.

The wood of the tree is yellowish, and is capable of a high polish. I believe it is the best wood in India for printers' blocks, &c.

This tree grows all over India; and, like the coffee plant, embraces some 30 degrees of latitude, and so is capable of bearing a great variety of climate and soil.

THE GUAVA TREE.

This excellent fruit tree could be easily introduced into America. The Southern States would be sufficiently temperate. It grows in climates in India colder than any part of Georgia. No tree yields a greater abundance of fruit, and guava jelly is known all over the world. These trees, &c., not being in America, shows a great lack of individual enterprise.

THE ORANGE TREE.

This tree could be introduced from the North East of India, from any part, from 25° North Lat. up to 33°. Of course there would be a vast difference between those trees from high latitudes, and those of West Indies under the line. My belief is, that the knowledge of the West Indies has been a misfortune to America, for all experiments in introducing exotics have invariably been made on West India plants, because they are easily to be obtained; and as such experiments have proved failures, ergo, none of these exotics would answer in America. There is the pomegranate tree, which thrives best in 15° of N. Lat. in East India, yet it is grown all over Georgia.

THE LIME TREE.

This tree is a kind of companion to the guava and orange trees. Surely this would do well in America. It yields most abundantly of fruit; and is all over East

India, under various degrees of climate, in the hot burning winds of the North West of India, or in the damp chill, and, in cold season, frosty climate of N. East of India. Or, if this will not satisfy Americans that it would do well in America, turn to the Russians, and ask them from the bark of what tree it is they make such an amount of cordage, and they will point out the lime tree.

THE CITRON TREE.

This is another companion to the above three trees, and is like the others in respect to soil and climate; and if other trees be evergreens, this may be called an ever-bearing fruit tree.

THE INDIAN GOOSEBERRY TREE.

This cannot be ranked among the tree tribe, nor hardly does it partake of the nature of a shrub. It is about two feet high, and yields a quantity of fruit. This gooseberry is a different fruit from that of Europe. It is not near so tart, but is used extensively for tarts. It is a pleasant fruit to eat, and, unlike the European gooseberry, there is no danger to be apprehended from its use. It grows wild in all the Eastern and N. Eastern parts of India, and in season the markets are well supplied with it. It is one of the fruits that are admitted at the dessert table in India. It is very easy to propagate it, to any extent.

THE NUTMEG TREE.

This has been found grown wild in Assam, lat. 27° 30″ and probably it would succeed in higher latitudes in America; all the north-eastern parts of India are intersected by the Himalaya range, and therefore the cold is greater than in the same same latitudes in the north-west. The Himalaya mountains dip five degrees southward from west to east side of India.

THE BAMBOO

Could successfully be introduced in America. It is of the greatest use in India, and is converted to various purposes. Walking-sticks, fishing rods, bed-frames, handles for tools, roofing for houses, sieves, baskets, hampers, &c., are made of it. It is also used as vessels for carrying water in, amongst the more savage races of India. It is made means of to hold rice, tobacco, salt, &c. Of it mats are made. It is pickled; and, though last, one of the best, perhaps one of its greatest uses, in a mercantile view, is the immense quantity of paper made of it. The soft parts of the bamboo are used for pickles, and as a vegetable. The young bamboos are used for making paper. The following is nearly the mode of manufacture:

The green bamboo is placed in a vat about ten feet square, (it may be larger,) built of brick and lime, and from three to four feet deep, and is allowed to soften in the water for several days. It is then taken out, and pounded until it becomes a pulp, from which the coarse and knotty parts, &c. are carefully separated. It is then mixed up with water to the consistency necessary for paper.*

The bamboo grows to a great length, from twenty to thirty feet, and the larger kind are thirteen to fifteen inches in circumference. The thickness of the shell in

* The Chinese make coffins of the bamboo, having no other wood.

such is about three-fourths of an inch, giving a cavity of three and one-third to four and one-third inches in diameter. These are the sizes generally of bamboo made use of as vessels to convey water, holding rice, tobacco, &c. They are cut in lengths, for water purposes, of $3\frac{1}{2}$ feet, and a stick is driven down to open a communication between the different chambers, so that the water may flow into them all—the bottom is always cut at a joint. A man or woman takes four or six pieces of such bamboo to the water side, and fills them, then connects one half to one end of a bamboo, and the other half to the other end; the bamboo is then hoisted on the shoulder, with these vessels filled with water hanging from each end and balancing each other. The Assamees and Tartars, when no other means is at hand, boil their rice in them, and the Nagas, at their salt springs, make use of them as substitutes for evaporation pans. Such are the uses of the bamboo.

THE CANE.

This useful article grows as a brier, and forms an impenetrable under-wood, generally forty to fifty feet long. The outer bark is like that of the brier, covered all over with spires. It is a bulwark against the wild elephant or buffalo, neither of whom can make his way through it. It generally is the haunt of tigers. We make use of it every hour in the day—we sit on it, sleep on it, and devote it to many other purposes. It is an article of great commercial value all over the world, and well worthy of attention. As stated, it is an under-wood; I never saw it grow out of the forest, and can say nothing of its cultivation, if it be cultivated. The cane is very various in its thickness, never, that I have seen, exceeding about four to four and a half inches in circum ference.

THE INDIAN RUBBER TREE

Grows to some seventy to ninety feet high. mmense forests of it are found on the west side of the Burampooter, extending along the Meeree and Abor mountains. It is a stately tree—it is said some are 100 feet high. The rubber from this tree has not answered for exporting from India. The expense of making is a mere trifle; but, whether it is owing to the tree having been accustomed to a colder climate, or from some chemical property in the rubber, it cannot bear the heat of a passage to Europe. It becomes a fluid during the voyage. Otherwise, in cold climates, it is equal to other rubbers.

THE KIA PUTTY, OR KAYAPOOTEE OIL TREE

Is also a very large tree, and yields its oil in the same manner as the pine tree does the turpentine. The oil is thick and is often used by itself for varnish over maps.

THE BLACK GUM VARNISH TREE

Is another very large tree. The gum is obtained in a fluid state, and remains so. The Tartars obtain it by making slits in the bark of the tree, and then, filling it into small pieces of bamboo, sell it for varnishing over handles of their dahs (swords). A wine-glass-full sells for 8 annas, or 25 cents. However, it is more generally bartered for rice, salt, &c. Some idea may be had of its adhesiveness from the fact that the Tartar dah is made use of, not only as a sword in war, but also in peace for felling trees or cutting up firewood. Consequently the friction on the handle is very great—notwithstanding, the gum adheres for years, except on the edge or end of the handle, where the wood itself becomes less or more worn. Even if these three last trees should not become of any great value as articles of commerce, they would be ornamental—and it would be difficult to say what the value really might be without knowing fully the quantities of milk, oil, or gum that each would yield, and of that I can give no information. It would be easy to obtain a few plants, or seeds, or nuts of them.

JOINT GRASSES.

This is not a grass that is generally known in India. I never met with it in the South, North-west, or in Bengal. It grows in the Tartar country; generally in the margins of forests, where there may not be too much shade; a forest being partially cleared, it springs up in places where it perhaps never existed before, or if it did, not for centuries past. The grass will run to a length of some fifteen feet, and will rise, if there be any support, five or six feet; if not, it will grow up some three or four feet by its own support. It is not a wiry grass—the joints are some six or eight inches long, with four or five blades of grass about the same length growing out from each joint. The joints near the ground are harder and brittle—those near the top soft and juicy—with a luxuriant termination of soft blades similar to those from each joint, but are softer and thicker. On my arrival in the country I found there were no cows, goats, or sheep. These I introduced, and at the same time a quantity of gram upon which sheep are fattened in East India. After the arrival of the cattle they declined the gram, and I found on inquiry that they had been browsing on this grass, and upon which they continued to feed. They all became as fat as if fed on the gram, (a kind of pulse,) which remained on hand, there being no use for it. This grass would be valuable in America, and should be introduced—for there is very little vegetable in the grass way in all America that I have seen; and indeed I believe it is frequently the case, down to the South especially, that a man may have from twenty to thirty cows and not a drop of milk for his breakfast. Of course there are two causes for that; the first, the worst of all, is idleness and bad management combined; the second is the scarcity of vegetation, or fodder.

Beyond the introduction of this grass, it will be well to suggest to Americans generally the great facilities to introduce a great variety of grains and pulses required in America, from other countries. I will here name some that seem to be worthy of attention, and may lead to further inquiry for others:

Of Cereals, there are a great variety of millets and small grains—and several kinds of wheat and rice.

Of Pulses, Indian gram and dholl would be valuable.

Of Oil Seeds, the sesamum, mustard-seed, &c. There are upwards of twenty articles from which the Indians extract oil.

ELEPHANT MUKNA.

The sheep of India would do better for the hotter parts of the United States, where wool could not be raised, than sheep of colder climates. Wool of Indian sheep is coarse, somewhat like an African's hair.

Camels would be very useful, and I believe would do well in America, at least in the Southern States. Elephants, except in wet, marshy countries, without roads, are not very useful. They are slow. Twenty-six miles would be a long day's journey for an elephant. However, both camels and elephants could be introduced when of a very small size.

Under this head I may, perhaps, show some kind remembrance of two pets I had in the far East.

The one was an elephant called "Mukna."* On my first arrival, the Governor of the North East frontier purchased two elephants for me. After the purchase he became aware that the larger of the two was a fierce animal, and had killed two men, and told me that such was the case. As soon as I saw the elephant he made a dart at me, to lay hold of me, which gave me an itching to tease him; which I generally did by pointing my finger, or a stick, at his face, always standing at a safe distance, with my friend well tied up. However, I never admitted the mahouts (keepers) to steal the elephants' rice, and always stood by to see them fed; and I believe they knew that was the case, for as soon as they saw me there was great excitement, and a sharp lookout for their food. Sometimes I would feed my friend, and then he could be docile to me; but otherwise, he invariably made

* Mukna is a general name for elephants that have not large tusks.

the same dart at me to lay hold of me. Being amongst a lawless people, I was frequently attacked by them in great numbers in travelling through the country; then my elephant stood my friend. When hemmed in on all sides, I placed Mukna, whom I rode, at the head of the other elephants, and bore down on the leader of the savages. The elephants took a pleasure in rushing on them in all their fury, with tails up and trumpet (proboscis) sounding. Mukna universally carried me helter skelter into their midst, and kept up pursuit. Once, about 150 men surrounded my elephants; I had six; they were all captured but Mukna, on whom I was riding some distance in advance. I had no arms but a sword, and was in some doubt what to do; whether to run for it or make an attack. However, I was not long considering the matter; one of the savages hurt the companion elephant of Mukna, and made it sound its trumpet; when Mukna rushed to the charge, regardless of rider or driver. The pikes of the attacking party I beat aside with my sword, and managed to reach the other elephants. Mukna made a rush, and scattered the captors of his companion; and with the two elephants I then faced the whole party, soon recovered the others, and the people were scattered everywhere. One of them gave Mukna a stab on the side; it made him trumpet, and I feared he was severely wounded. I gave chase to the man, feeling every resolve to punish him if I could get at him. He escaped into the village and into one of the houses. Mukna soon laid hold of the wood frame work, and shook it down; and one by one, as the elephants arrived, they commenced at the same work, until the whole village was destroyed. At last, through the aid of Mukna, I succeeded in making the savages have some respect for me. Sometime after this occurrence, I had been teasing him after feeding him; some ryots (farmers) came up to speak to me about some lands; I turned my back on the elephant, (he was tied up, and could not reach me). Soon after I got a stunning blow, and thought the ryots made an attack on me. As soon as I recovered, I turned round on my supposed assailants, when, to my surprise, I found Mukna had taken up a long piece of bamboo, and whirled it round with his proboscis, and struck me along the side of the head. He seemed to be quite delighted, and carefully laid his bamboo along side of him, and when I attempted to approach, he laid his proboscis on it, as much as to say, *here it is.* One day while I was travelling over some rugged hills, 150 to 300 feet high, a large tree fell across the pathway; Mukna got across it at his ease; he was the largest elephant in the country. However, when his companion came to it, she managed to get the fore legs over, but her hind ones she could not raise, nor could she raise the fore legs to go back. She was in despair of working herself out of the predicament, and trumpeted; Mukna at once turned back, and at my desire the driver left him to himself. He got over the tree again, placed his head to the rump of his companion, and raised her hind part over the trunk of the tree. After many and faithful services, and narrow escapes, poor Mukna was poisoned by the Tartars. I let him loose one morning after feeding him; a while afterwards he returned to the house, and came to the door and gave a low trumpet; I went out, and he walked up quietly to me; I laid my hand on his head, I could not make out what was amiss with him. He laid down, and fifteen minutes afterwards was dead. I buried him deep where he died, sooner than have the poor animal dragged by the other elephants; but in a few days I was obliged to decamp from the fumes exuding from the earth. Some ten months afterwards I sent a party to take up his bones for a skeleton, but there was even then so much of the flesh undecayed, I was obliged to give up the work. A few months after I was attacked by the Tartars, and several of my people killed, as already shown.

Another extraordinary pet I had was a blue bird called "Porphiro," and in that country "Kiam." Some of the same kind are in the Washington Museum. Kiam acted a faithful watch-keeper, and in the early mornings used to get into bed with me, and in getting under the clothes, would take a nap. He was a great favorite with the servants, and from his curious strutting, used to afford them a great deal of amusement. I mention these matters simply to show a few traits of the elephant and of the kiam; and would wish to dwell upon this subject, but space will not admit of my doing so.

The feathered tribe, that is, the game kind, are very numerous. Oolongs are perhaps the least known; they are very shy, and seldom or ever shot. They are as gross as geese in body, and stand in height, four feet. There are many kinds of fowls; the species of duck cannot be less than a score. Most of the pheasants could be domesticated. The flesh of the oolong is the finest flavored of any of the birds I have met with.

Fish is almost as numerous as insects in India. Every small pool of water contains fish of an excellent kind, and weighing two to five pounds' weight. It is surprising to find fishes where they are to be found in India. I have seen the natives in the North East of India, both to my surprise and amazement, dig fish out of the earth. The fish is called "earth fish," "Zeemen ka mutchee," of about five to seven inches in length, flat, and black in color, flesh hard, and in flavor somewhat like an eel.

I will give here, in hopes some enterprising gentleman may try the experiment, an account of the manner of the Chinese hatching fish, from Mr. Martin's work.

"Hatching eggs by artificial heat is well known and extensively practiced in China; as is, also, the hatching of fish. The sale of spawn for this purpose forms an important branch of trade in China. The fisherman collects with care on the margin and surface of water, all the gelatinous matters that contain spawn of fish, which is then placed in an egg shell, which has been fresh emptied, through a small hole, which is then stopped, and the shell is then placed under a setting fowl. In a few days, the Chinese break the shell in warm water (warmed by the sun); the young fish are then kept in water until they are large enough to be placed in a pond. This plan, in some measure, counteracts the great destruction of spawn by troll-nets, which have caused the extinction of many fisheries."

This art carried out would be most valuable to all countries, and would be a means of making the water equally, if not more, productive for means of support than land; for, by such art, every piece of water might be filled with fish.

The Opium Trade.

In entering on this subject, I fear I shall have to represent a picture of misery more extensive and more astounding, than has ever been heretofore set forth in the annals of history. Were it a trade in human flesh, it might be to redeem the negro from the darkness of paganism, and to establish him through Christianity in the brightness of Eternal Glory. It might be the means of redeeming him from being sacrificed at a feast or a wedding to the atrocity of his capturer, or being quartered before some hideous figure representing some monster, or some monstrous idea of the Creator. All these services might spring and do spring from slavery, but it is not a good way to effect reform. Still slavery could point to many fair and bright spots, to redeem it in a manner in the minds of the justly reflecting man. But opium trade—alas, there is no one bright spot, no one redeeming virtue can be found in the whole course of its man-devouring, hellish course. You, oh Americans! would rend your States into petty governments and principalities, upon the subject of some three millions of slaves, well clothed, well fed, and converted from paganism—simply because they bear the name of "slaves." But here, this ruffian trade sends more than that number yearly to a premature grave, and destroys in man's loins that procreative seed, from which would spring into birth a great number more.

England, whatever my attachment may be to some of your citizens individually, I must here represent yours as one of the worst governments that ever cursed the human race. And I here state before God and man, that you murder and destroy more people than all the governments or nations over the face of the earth put together; that your government is a stigma and shame to every white man, and that it is the scourge of the weak and powerless—a government, oh, England! that is bringing a curse upon yourselves.

The first assertion that an apologist for this trade would make, and which England makes, is that China is satisfied, or they would not buy opium. I will simply give the following memorandum of the Chinese government acts to save their people, and all may then judge of its truthfulness; and it might as well be said, that the father was satisfied with the gambling-house, the brothel, and grog shop, because his infatuated son, over whom he had lost all control, frequented them.

In 1800, the Emperor of China prohibited the importation of opium into the Empire, and death and confiscation of property were decreed against all retailers or cultivators of it.

In 1809, under the Emperor Kea-king, the Hong merchants were required to give bonds of security, that all ships discharging cargo had no opium on board.

In 1815 the Emperor directed that the laws should be rigorously enforced against natives dealing in opium.

In 1820 a prohibitory proclamation was issued against opium.

In 1831 another law was enacted, to flog and transport those who refused to point out the seller of opium.

In 1832 an order was again issued against the importation of opium.

In 1834 the order was re-published.

In 1837 an order was issued to send away to their own country all "opium warehousing ships." The same order was issued again in the same year.

In 1838 a China man was sentenced and strangled in face of the English factors at Canton, for trading in opium.

In 1839 full power was given to Commissioner Lin, to suppress the opium smuggling.

In 1849 20,238 cnests of opium were forcibly taken from the English, who then signed a bond in which they solemly bound themselves for ever, not to introduce opium into China. The whole of the above opium was, before English witnesses, mixed up with lime, salt, and water, and so destroyed; its value, some $6,000,000, valuing at $300 per chest.

For this noble act of the Chinese against smugglers, the English government declared an unholy war against China, of which all know the history. But have any reflected on that war, made on a virtuous government, at least upon a government that was acting nobly in defence and protection of its subjects from demoralizing contaminations? England made war on China, because she dared to attempt to put down smuggling into her own country.

That is the country that cheers a Kossuth, for making

war against the wrongs of Austria. That is the country, on the other hand, that protects the violators of the laws of nations, the laws of humanity, and the dictates of honesty. She supports Kossuth by her empty cheers, because it creates a little political capital; it will give Austria, &c. more to do at home, and she will be the more free to carry on as she pleases abroad; and she would also ensnare America into hostility to Austria and Russia, and forsooth all that is—humanity! Oh, Americans, look to China, and see some three millions of souls dropping year after year into the grave, the victims of opium. And, behold the price of England's inhumanity for the wholesale destruction of that hecatomb, some 45,000,000 dollars yearly—that is the price of the blood of millions of Chinese destroyed year after year, by some 60,000 to 70,000 chests of opium, or 8,010,000 lbs. to 9,345,000 lbs. of a deadly poison.

I will insert the progress of the increase of opium, and will afterwards show that the more it yields the British East India Company the more that Government squeezes all parties; for that Government derives all the profit, with the exception of that which is made by the smugglers, its tools, which is of no great amount comparatively, and is divided between a few mercantile houses in China. Yardine, Matheson, & Co., the principal purchasers, are supposed to have realized £3,900,000 in twenty years. The other six houses concerned in tho trade probably may make twice as much more. It would give £9,000,000 in twenty years. This sum would be only equal to one year's sale. Opium shipped to China from East India for following years is :—

	Bengal Opium. Chests.	Malwa Opium. Chests.
1820,	3,591	2,278
'30,	7,443	12,856
'35,	14,851	12,933
'40,	18,965	18,321
'45,	21,457	20,660
'46,	20,000	19,063
'47,	21,650	20,523
'48,	28,000	17,490
'49,	36,000	18,532

But this is not the whole of the opium; it is only a part of that from Malwa, which passes through Bombay under a *pass* from Government (British East India Company), and that which is sold at Calcutta for the China market, and for which express stipulations are made by the British E. I. Government that the opium is to be shipped for China; because, if retained in India and sold there, it would be interfering with another abomination of theirs, to raise money, viz.—the establishment of opium smoking shops in their own dominions. To the keepers of these shops Government retails any quantity of opium at more profitable rates.

Nor are the above amounts, as returned, correct, as may be seen from the evidence before the Parliamentary Committee of 1847, on commercial relations with China, of F. W. Prideaux, Examiner of the East India House, in which he stated that the amount of Bombay opium for 1844–45, was 29,593 chests, instead of 20,660 given in the above table; and for 1845–46, it was stated by opium officers that there were 30,000 chests, instead of 19,063 returned; therefore, it is pretty clear that the true amount of that drug is concealed. However, it is shown by such returns that 54,532 chests of opium, containing one pecul, or 133⅓ lbs. each, were sold in China in 1849. Say price in Calcutta 1100 rupees, or 550 dollars, sale price in China say 600 dollars—32,719,200 dollars.

The import of opium into China is shown to be more, by the evidence of Government itself, than is stated in the above returns, taken from the tables of the Parliamentary Committee, and from the "China Overland Mail." Beyond that amount, there is that of the consumption in India. The East India Company's Examiner says, in his evidence: "The value of the opium exported from India in each year (1844 and '45) must be from five to six millions sterling." It is now increased, at the lowest calculation, 100 per cent. so it may be said it now reaches £10,000,000 to £12,000,000 sterling. And there is the secret that neither honesty, humanity, nor the most solemn treaties, has power to restrain the cupidity of a Government, placed over 200,000,000 inhabitants, and that awes some 500,000,000 of Chinese into submission to the most unjust trade that ever disgraced any nation.

THE EVILS OF OPIUM.

If I were simply to give my own evidence on the subject, it might be held in suspicion; or, were I only to give the evidence of parties opposed to the trade, it might be said it was only a morbid sensibility of a party desirous to put down the interest of another party whom, in their enmity, they call "smugglers." I will state a few instances of the evil of this drug that I have personally witnessed.

The first and greatest evil is, that it exterminates the human race. Opium is put down, used moderately, as aphrodisiacal, and hence it is reasoned the cause of the wretched women who assemble round the opium smoking shop. Of this I have no knowledge; but there is no disease in the North-east of India so dreadful as that for which it is said to be a cure, or is so very common amongst the people. The opium eater is a lost man, from his first initiation into the habit of smoking or drinking opium. If he commences early in life, he never marries—if he commences at the time of his marriage, he never has more than one or two children—if he commences it one or two years before marriage, it is rarely there is ever a child, and if there be a child, it is owing to a very moderate use having been made of the opium, arising from the want of means to purchase it, or other causes that may place it out of reach; therefore, opium is detrimental, in the first place, to the propagation of the human species. In the next place, those who indulge freely in its use live but a short time.

It is melancholy to travel through a country where opium is used; in every part there is the foot-prints of man; there are the few orange trees and guava trees, &c. where once was the cultivated garden, and no trace otherwise of man than the rich weeds and grass that spring up from the land manured by a decayed fallen house. In the places where villages were, are marks where houses had been; others partly fallen in, and green fungi growing all about. The poultry, now wild, still haunt round the place. Again, another village may be met with a few solitary houses remaining, with some twenty to thirty women, two to five children, varying from one year of age to ten, a few young girls, and the remainder women, varying from thirty to fifty years of age—all miserable and squalid—and perhaps there might be one or two males. In other villages again, there are more inhabitants and new houses added, but it is owing to the remnants of other villages resorting there from the ruins of their own, and the advance of the forest. This is the condition to which opium reduces a country, and, as in the valley of Assam, so it is in the Tartar country to nearly an equal degree of desolation. Such has been the result of the use of opium as witnessed by myself. I will now give the experience of parties interested in the trade or otherwise.

Mr. Martin, one of her Majesty's treasurers in China, represents the use of opium as follows: "The continued action of opium as a sensual stimulant tends rapidly to the wasting of youth, health, strength, and beauty; those who begin its use at twenty may expect to die at thirty years of age. The countenance becomes pallid; the eyes assume a wild brightness, and memory fails, the gait totters, mental exertions and moral courage sink, and frightful marasmus or atrophy reduces the victim to a ghastly spectacle, who has ceased to live before he has ceased to exist."

W. Hamilton Lindsay, Esq. M. P., says: "As it is, nothing can be more injurious to the British character than the mode in which the opium trade is at present conducted. It is now real smuggling accompanied by all its worst features of violence."

Captain Elliott, late her Majesty's superintendent in China, says: "It is intensely mischievous to every branch of the trade; that it is rapidly staining the British character with deep disgrace."

Sir John Hobhouse, President of the Board of Control, said in Parliament: "he could not but deprecate it as a vice, for a great vice it was."

Lord Sandon said: "It is a disgrace to a Christian country, to carry on the opium trade as we have done."

Mr. Squire, agent of the Church Missionary Society, said of the opium shops: "Never, perhaps, was there a nearer approach to hell upon earth, than within the precincts of these vile hovels. Truly it is an engine in Satan's hands, and a powerful one; but let it never be forgotten that a nation professing Christianity supplies the means; and further, that that nation is England."

Rev. Howard Malcom, of the United States, said: "The great blot on foreigners at Canton, though not all, is the opium trade. No person can describe the horrors of the opium trade. That the government of British India should be the prime abettors of this abominable traffic is one of the great wonders of the nineteenth century. The proud escutcheon of the nation that declaims against the slave is thus made to bear a blot broader and darker than any other in the Christian world."

I will give the following extracts from a table given by Mr. Martin in his work on China, to show the number of smokers. Mr. Martin has made up the table to 1835; I will continue it up to 1849:

	Total chests of Opium.	Total candareens.	Smokers, at 3 cands. per day.
1820	4,287	400,440,000	365,699
1823	5,073	505,000,000	461,187
1826	8,452	894,160,000	816,584
1829	11,080	1,132,800,000	1,034,520
1832	15,662	1,615,920,000	1,475,726
1835	21,677	2,233,800,000	2,039,998
1840	42,117	4,195,760,000	3,831,744

	Total chests of Opium.	Total candareens.	Smokers, at 3 cands. per day.
1845	- 37,286	- 3,325,720,000	- 3,037,190
1846	- 39,063	- 3,887,560,000	- 3,550,283
1847	- 42,179	- 4,194,760,000	- 3,830,000
1848	- 45,490	- 4,338,800,000	- 3,971,507
1849	- 54,532	- 5,103,840,000	- 4,661,041

I have followed this calculation in the same manner as Mr. Martin did, viz: one chest of opium contains 100 catties. The Patna and Benares opium contained 800 candareens per cattie, of pure extract, at fifty touch, and the Malwa opium, 1200 candareens per cattie, of pure extract, at seventy-five touch. Three candareens are equal to 17⅗ grains—the quantity each opium smoker consumes per diem.

Therefore, the above table shows, to use the expression of Mr. Lay, one of her Majesty's consuls, that England, humane, liberal England, hamstrings 4,661,041 Chinese yearly, and when we consider that these only live some seven or eight years,—the 3,831,744 that were the consumers of opium in 1840 having dropped off by the year 1850—and the 3,037,190 of 1845, will drop off by 1855, and each succeeding year millions will drop off, and infatuated millions will fill up their place. Oh that China was become bankrupt that she might be saved!

It is hard to give credence to this, that any Christian nation could, for the sake of money, be such wholesale butchers; but, above are the figures; *give any man in this country at the rate of 17⅗ grains of pure extract of opium a day and how long will he live?* It, of course, does not follow that every man who uses this drug takes 17⅗ grains in the day; at the commencement it is always less, but, to the end, it is more than that. No smoker of a year or two standing, will do aught in the morning until he either gets his opium pipe, or his opium diluted in water to drink. And when he is in the habit of using it for three or four years, he has no power over his body, and his spirits are in that state of depression, that a man on the moment of preparing for public execution could not be more mentally prostrate.

Is this generous England, so intermeddling in the domestic affairs of foreign nations, prying into their prisons, to get some loop-hole to make an empty parade of her good will towards the oppressed? Alas, there is nothing in all that but to blindfold. *Think that a nation that can for some* 45,000,000 *of dollars of blood-money send yearly a hecatomb of three to four millions of souls prematurely to their grave, is humane!* Nor does this accursed poisoning trade stop in China only! Alas, no; England has nearly 200,000,000 of subjects in East India, amongst whom she is pushing the same vile trade, and viler and more basely still in stooping to retail this drug of hell by pounds and half pounds, to smoking shop-keepers, licensed by them, and that never existed before this morbid craving after money suggested it to them, so that in East India the government are the tempters of the people into a vice they say they are not accountable for in China. China, India, Java, and all eastern countries, will be exterminated by a few millions of people styling themselves Anglo-Saxons, because nations and people will not inquire into the state and condition of that people, who are beginning even now to waste away from the face of the earth themselves—as is seen in the last decade, although the starving Irish raised the population of most of her large towns in England, as that of Liverpool for instance, to 50 per cent. shows only a miserable increase of 11 per cent. which is made up of Irish and other races. That country, which is existing upon the prestige of her name earned for her by the unfortunate, crushed Irish, is now ready to topple over herself. She has no more Irish to fall back on in case of emergency; in the last decade they gloated in the extermination of millions of that generous people; she, through her blinded bigotry, insulted the suffering remnant by malignant penal laws to deprive the people's clergy, not of their titles only, but of the power to ordain further any clergy at all—she dare not put that act into execution. Let her now meet the enemy in the field, and she will find that she has dried up her resources, that she has by her injustice and cruelty, scorpion-like, turned her own sting on herself—and I believe the remnant of the Irish yet in Ireland and in England, and descendants of Irish, would be found to equal the whole number of that Saxon race that is now working in the world so fearful an amount of wrongs, and that the Celtic race all over the world, is as two to one of the Saxon—but more of this at another time.

EFFECT OF OPIUM ON TRADE.

If merchants will consider China in her greatness, and cast away all that prejudice arising from the misrepresentations of English writers whose policy is ever to blacken the characters of those they oppress. If a murder be committed in Ireland by an unfortunate, outraged tenant, who, with his wife and family, is driven from house and home into the ditch at the road-side; if that unfortunate man, maddened to see all that is dear to his heart famishing in that ditch, be driven to a step of dire revenge, and shoots the author of his wrongs, it is not that miserable wretch who commits the murder, oh no, it is the IRISH! Of the numerous murders committed in England we hear little—but, when they are commented on, it is the individual who committed the murder, and not the English. In like manner, are the Chinese misrepresented. But for an account of that people I appeal to the French nation, who are alone acquainted with their histories and writings, and have a friendly intercourse with them. There are none of the Eastern races so docile, so cleanly, and industrious as the Chinese—and none carry improvements so far, wherever they migrate. Rarely, or ever, is a Chinaman found, of the great number of them in Calcutta, before a criminal court of justice; and even before they would take another before the courts, they would forgive him the offence he committed against them, or rescind the debt that might be due to them, although they are but shoe-makers generally, who are in that city.

The population of China in 1812 was 361,000,000 of inhabitants. Say the increase since in each decade is ten per cent. and say four decades, or forty years, up to 1852, it would give her 537,320,700 of souls. They require clothing of all kinds, cotton and woollens; they require rice, wheat, flour, &c. They hardly have any timber to pack even their teas in, or to build the myriads of junks employed in carrying salt and grain for that population; even frequently their coffins are made of bamboo work. But China is paralysed by the drain of some 45,000,000 yearly of dollars for poison, and which reduces millions of her people to be a burden on her. This traffic "hamstrings" the country, and stirs up the people to a hatred of all trading nations; and this will be fully shown in her treaties with England and America, compared with that of France, and her manner of receiving the French representative, to whom she granted all he asked, even to full religious toleration. If that drain for opium was allowed yearly to circulate in a useful trade, how far more England would benefit, and to how far greater extent would American cotton be consumed by China, either in English piece goods, or in American coarser cloths. Taking the product of the loom from either parties, would be equally consuming the American raw material. What does America now gain by allowing a few individual Americans to participate in some paltry profits in that trade, and prostitute the American flag to the vile purposes of smuggling, in comparison to what she would gain if all parties acted fairly and honorably with China?

England makes treaties with China, and pays them no respect; but England regards her treaties in proportion to the strength of parties to compel her to do so. She made a treaty with China not to import opium. The British East India Company, which is controlled by her Majesty's Government, grows the opium purposely for China, and with the full knowledge that it is to be smuggled. The English flag protects the smugglers; but they have not stopped there. The English Government has, in Hong Kong, licensed, in opposition to the laws of China, and in violation of her treaty, opium smoking shops. This is the quasi liberator of the oppressed of Europe, the Don Quixotte of the discontented, in the hopes to give other Governments enough to do at home in suppressing disturbance, so as to prevent them looking after her affairs over the world, or to the misery of the Irish. That is her policy. She would crush every spark of liberty to-morrow on the continent of Europe, as well as poison the Chinese, if it tended to her aggrandizement, or extension of her trade.

Let Americans, in the name of God, do all in their power to discourage this nefarious trade, and adhere to their treaty with China, *and compel every citizen to do so.* If America will do this, England, from shame, will be obliged to suppress this trade, and a healthy commerce will then spring up with China. Probably America may, from her western coasts, open a trade in rice, grain, timber, and cotton, and of ten-fold more consideration than it now is with China, or a trade altogether greater in itself, than that now with Great Britain. Would not this be better, yea, would not the sacrifice of all trade be better, than that any of her citizens should be permitted to violate the solemn treaties entered into, and which she is solemnly bound to maintain in full force and spirit? No American should be allowed to encourage and abet, or aid or participate, in the trade, in any way whatever. Is it better that one or two individuals should make a few hundreds of thousands of dollars, to the detriment of American honor, than that the whole country should derive the benefit of an extensive commerce and a free intercourse with so many millions of people?

I will mention an article, of seemingly very trifling consideration, with which America might at any time have supplied China, viz., a tea-chest. The boards could be cut and prepared to be put together in China, and conveyed from America. The bottoms, lids, sides, and ends of each box should be carefully placed with each other, so that no confusion would arise, and that the boards or board of one box might not get mixed with the other. The box should be made of the lightest timber, no matter how soft, or inferior the wood, but pine wood, or any wood having a strong smell, would not be suitable, as it would injure the flavor of the tea. China exports 120,000,000 lbs. I believe her boxes, the largest, do not hold much more than half a pecul, 66⅔ lbs. to 80 lbs. Therefore, she requires nearly 2,000,000 of chests at that dimension; but as the tea is packed in boxes containing five cattie chests, ten cattie, &c., &c. to the largest size, there would be some 3,000,000 of chests required—and for domestic use, she would require several times that amount. America, judiciously entering on this one branch of business, and carrying to China tea-chests of all sizes, packed up in boards, which would not occupy much storage, would make ten times as much as she can by any connection through a few individuals in the infamous trade of opium. And if there be this opening in a branch of trade heretofore unthought of, what must be the amount of trade that may be carried on with 500,000,000 of people. Let America act honorably and conscientiously.

Americans, your glorious stars and stripes are become the inglorious flag of smugglers; your solemn treaty is torn into shreds; your faith is violated; your colors float, and your eagle spreads his wings from the mast heads of seven opium smugglers: four on the coast of China, and three between China and East India; and through the agency of a few, the ensigns of your country are become the scape-goats of opium monopolists and English cupidity! Look to this *pour amour de la patrie.*

The drain of silver from China by this opium, is striking at the root of commerce; and notwithstanding the large amount of cash America and England pay for their goods beyond the value of the goods exported, which may be now some 20,000,000 of dollars, (in 1845 it was 16,000,000 ready cash,) and as the import into America of tea for 1851 is upwards of 28,700,000 lbs., ready cash payments must have greatly increased on the part of America; therefore, there is no way to increase the trade with China but by putting a stop to the opium drain, which is so much more serious than is represented. Next year (1852) the charter of the British East India Company is to be brought before Parliament, and it is the more necessary to conceal the amount of this fearful drug that is produced; therefore, the drain for opium on China, yearly, cannot be less than 44,000,000 of dollars, and the destruction of some five to six millions of people annually.

In proof of this injury to the trade of China, I will not fill up this article further by the testimony of individuals, but will give two extracts from the deliberate judgment of a body of Englishmen, the Parliamentary Committee of 1847; and since their report was made, the evil has increased 100 per cent.: "The payment for opium, as will be seen from the inordinate desire for it which prevails, and from the unrecognized nature of the transaction, which requires a prompt settlement of accounts, absorbs the silver, to the great inconvenience of the general traffic of the Chinese." And again: "Opium trade, however, already flourishes at

Fouchoo-foo, with all its demoralizing influences on the population, and embarrassing effects on the monetary condition of the place."

There is a very erroneous impression, viz. : that the opium eater is not a dangerous man under its effects. The facts are quite the reverse. On all war or exterminating excursions of the Singphoos, they keep up the excitement by opium. Under its influence, the face becomes fearfully puffed, the eyes dazzling, and then the person so affected is roused by the least provocation to perpetrate any murderous act; and so dangerous are they under this excitement, that it is necessary to sooth them. When the English goverment established smoking shops at Hong Kong, they were obliged to prohibit the smokers to carry arms into the shops, under a very heavy penalty of fifty dollars. It is only when the effects die off that the smoker becomes quiet, because his body and mind are then prostrated, and he trembles from head to foot.

The English Government is not content with the word smuggling, they do not like the name; it does not sound well; and what would that Government do to wash out that foul stain? They would add to the villany of smuggling and poisoning, the crime of corruption of as noble an emperor as ever sat on a throne. They strive to corrupt him, that they might have the unfortunate Chinese at their feet, and then, instead of administering 54,000 to 80,000 chests of poison to the unfortunate people, they could send hundreds of thousands of chests, and destroy every good quality amongst them, and then say to them as one of the Governor Generals of India said to the native princes of that country: "You are incapable of managing yourselves; you are lawless, and we must manage you; you must receive into China our contingent forces. We cannot trust you to pay these forces, we must see them paid; therefore, allot to us the province of Quang-Tong."

The arrangement made, some squabble would soon occur between the English province and the Chinese. It would be magnified into insult, and that the people were tumultuous; extra forces and extra expense were necessary to curb them, and another province must go to England for that purpose. Next step would be to spur on the unfortunate Government into hostility. War would be declared, and England would be triumphant; the Emperor a prisoner, and declared dethroned, and some worthless member put on the guddee (throne). He would show his inability to reign, and would have to accept a pension from England. And his Excellency, my Lord Duke, as Governor-General of the Celestial Empire, would issue his ukase, and the world then might know the value of some hundred thousand chests of opium, and England's feigned love of liberty.

The amount of opium sales is now, say $45,000,000. Well, said the English, "your Majesty has not power to prevent us smuggling this opium, therefore, why resist? Put a duty on the opium, it will put money into your coffers." Here was the ruffian duplicity of that offer. Say 25 per cent duty was put on, it would make 11,000,000 dollars of a bribe for his Majesty. Then all over China, as well as on the sea-board, the poison would flow like a desolating flame; 200,000 or 300,000 chests would find consumers. His Majesty's revenue would be gradually increasing from $11,000,000 to $41,250,000, yearly. Was there ever a more insinuating bribe offered to any man by any villain?

Oh, Americans! give full credit to other races when they show a generosity to which we are strangers. The truly great, noble and high minded Taou-Kwang, present Ruler of China, and worthy Emperor of 500,000,000 of people, with true greatness responded: "It is true I cannot prevent the introduction of the flowing poison. Gain-seeking and corrupt men will, for profit and sensuality, defeat my wishes; but nothing will induce me to derive a revenue from the vice and misery of my people."

PRESENT AND FUTURE

OF THE

UNITED STATES.

If this subject be considered fully, and such reasonable data made use of as land-marks in guiding the mind, not to possibilities, but to probabilities, it may be made most interesting and useful; and may afford some faint glimmering light into the futurity of, not only the United States, but America generally.

	Acres.
The area of the United States, as set forth in table, is	2,081,759,000
The area of China, in miles, is 1,297,999, or acres, 830,829,200. China has more than half its area occupied by mountains.* Official returns of land under cultivation is 141,119,347, which, allowing one half to be under mountains, then two thirds of the arable lands are yet to be cultivated, - - - - - - - -	830,829,100
The area of France, 52,760,299 hectares, or acres,	130,391,040
The area of England and Wales, - - -	36,999,680
The area of Ireland, - - - - -	20,399,360

* It said by authors that two thirds is covered with mountains.

POPULATIONS.

Of America,	1850, - - - - -	23,267,498
Of China,	1812, - - - - -	361,221,900
Of France, say	1850, - - - - -	40,097,056
Of England and Wales,	1850, - - - - -	17,605,831
Of Ireland, say,	1840, - - - - -	8,175,124

NUMBER OF PERSONS TO EACH ACRE.

United States of America has -	1 person	to every	$89\frac{1}{2}$	acres.
China had a population in 1812 of 361,000,000, which say increased at a rate of 10 per cent. every ten years, would give, in 1850, 519,249,255, or -	1 "	to about	$1\frac{3}{5}$	"
or $2\frac{79}{141}$ persons to every acre reported to be under cultivation.*				
France, - - - - -	1 "	to every	$3\frac{1}{4}$	"
England and Wales, - - -	1 "	"	$2\frac{2}{3}$	"
Ireland, in 1840, - - -	1 "	"	$2\frac{1}{2}$	"

* It is surprising how fearful writers on China subjects are to give China her full complement of inhabitants. Gutzlaff doubted not that the population of China in 1812 was 361,000,000. Mr. Martin enters at length on the subject, to prove there were 361,000,000. Now, there is not one objection raised but can be met and set aside. Taking the population at the lowest point stated by objectors, and calculating it up to the present at a moderate rate of increase, and it will make it up to the amount of 361,000,000 in 1812. Martin himself, when he states the population to be, when he wrote his book (printed in 1847), 400,000,000, puts a dash after the number. If we whites do not come up to a great population like other people, it is because we are a turbulent race, ever desirous of shedding blood, and tyrannizing over each other. Where there is that degree of civilization, that there is in China, and without any severe, or in fact any, affliction or calamity having befel the Chinese since 1812, except the drain on the population by the Christian's poisoning them with opium; therefore, calculating the increase of the Chinese population at the low rate of ten per cent., it would be 529,000,000 in 1850.

INCREASE OF POPULATION.

United States,	Whites, -	from 1840 to 1850,	36⅓	per cent.
"	Free colored,	" "	14	"
"	Slaves, -	" "	27⅓	"
	Total increase, - - - -		$38\frac{6}{17}$	"
China, supposed to be at least, say - - -			10	"
France, for 14 years, 23⅘ per cent., or 10 years, -			17½	"
England and Wales, during 40 years, averaged 10 years, - - - - - - - -			15½	"

Ireland is no criterion to go by because of her emigration and reverses; therefore, to obtain a population in proportion equal to China—

The United States require - - - -	1,301,109,975
" " " to France, in 1850, say	640,541,230
" " " to England and Wales,	780,659,625
" " " to Ireland, in 1840, -	834,506,000

The two next considerations are:—First, Is the soil of America throughout equally capable of supporting the same population as the countries mentioned? China cannot have less than one third her area occupied by mountains and barren lands. A portion of England and Wales is also occupied by barren tracts and hills. Ireland has also a large space under hills or mountains; and although her soil is the richest of all, yet a portion of Connaught, as well as the north, is poor. American soil may, perhaps, be pronounced poor by some; but the American farmers are themselves the poorest—I do not mean in circumstances, but the poorest in science of cultivation and treatment of the land; I mean the Southerners, particularly. A portion of France is very poor.

Therefore, it may be inferred that the United States are capable of supporting at least some 700,000,000 of people.

Now, there may be those who say, " that amount can never be obtained; that it never had been obtained." It may not be amiss to draw the attention of the doubtful to the possibility of the States arriving at so extraordinary a degree of populosity. Ancient Egypt is represented as about 200 leagues, or say 600 miles, in length, and confined between hedges of mountains on each side of the hill, which, on the east side, approach that river to within a half a day's journey, say 15 to 16 miles. On the west side, the country extends from 70 to 90 miles; and from Alexandria to Damietta, the widest part, it is only about 150 miles; not more than one fourth the area of Texas. Yet it is represented to have contained, under the reign of Amasis, 20,000 inhabited cities, one of which—Thebes—is said to have been able to send forth at once from her 100 gates, 200 chariots and 10,000 fighting men from each gate, or, in all, 1,000,000 soldiers. Allowing, at a very low calculation, five inhabitants to every soldier, it would make the city of Thebes, one only in 20,000, to contain 5,000,000 of inhabitants. These historical statements of Egypt may be, perhaps, exaggerated, but could not be so far from the facts as to deceive all. We are rather prone to reduce populations below what they actually number. Before the first census of Ireland, taken in 1812, the English always represented her at less than one half of her numbers. The wonderful works of Ancient Egypt are certain evidence of an amount of population not to be equalled to-day by any country, even China comparatively. Probably, at the time of building the great wall, China might have an equally dense population.

To proceed further, and to see when the amount of 700,000,000 of inhabitants is to be realized. Table A will show the progressive increase of the people of the United States, up to the year 2,000, or 150 years hence. I have made the calculations at the rate of 35 per cent., nearly what it has been from '40 to '50, for the next ten years, to 1860; for the following ten, to 1870, at 30 per cent.; and from that period, at the rate of 25 per cent., for whites. For slaves, at 23 per cent. only from 1850; and for free colored at 15 per cent. These rates are very moderate compared with past years. And it may be supposed that 25 per cent. is not in keeping with the past ratio of increase, but that increase cannot be maintained. The same amount of immigrants may flow into America, but they cannot swell up the per centage to the same degree upon a large population as they could on a small one. For instance—500,000 on 10,000,000 would be 5 per cent., on 30,000,000 it would be only 1$\frac{2}{3}$ per cent., and so on.

To prove further that 25 per cent. is not much below the future increase, after another 20 years, it is only necessary to point out the following. The white population in 1830 was only 10,526,246. From 1820 to '30, the immigration was 140,000; and in 1840 the whites were 14,189,108; the immigration that ten years was 601,000. Showing that, deducting immigration and their increase, would leave the settled population's increase to be 28$\frac{3}{4}$ per cent., the increase of both to be 34$\frac{7}{10}$ per cent., as per census. Therefore, considering that the increase on large populations never keeps pace with that on smaller populations, the calculation of 25 per cent. is not under what the increase may be in some years hence; and if it be under it for the next 100 years, it will be as certainly over it for the following 50 years of the tabular calculation annexed.

The free colored population will most likely not keep its place as a class. The proportion of white males and females will be for the future more equal than it has been heretofore; and probably the female will preponderate over the male population; and, consequently, there will be more marriages, and less intercourse between the two races, between whom there is such wide difference, as an Irishman would say, in flesh and blood. The last ten years their increase has been only 14 per cent. I have calculated it, all through, at 15 per cent.

SLAVERY.

The slaves I have calculated throughout from 1850 at 23 per cent. From '30 to '40 it was $23\frac{1}{3}$ per cent.; from '40 to '50, $27\frac{5}{8}$ per cent. If they do increase at the rate of $27\frac{5}{8}$ per cent, they would be likely to gain numerically after some years on the white population. However, taking it at 23 per cent., it will be seen that the numbers of slaves, or negroes, will become excessively great. Their increase is much less likely to be interrupted than that of the whites. As slaves, they will be kept at home, cared for, and suffer no diminution in war or on sea. Upon the white population, all that drain of men will fall, as well as the drain from dissipation and intemperance. Therefore, it would not be surprising if the negro population do actually gain on the whites, and instead of being one to nine of the white population 150 years hence, they may be more seriously numerous, especially if they increase at the rate of 27 to 30* per cent. in every 10 years. It is, therefore, for the whites seriously to consider whether it would be for the interest of their race, and of the whole Continent of America, to overrun it with a negro population, which bids so fair, in the long run, to be so numerous.

However, the loss of some 3,179,589 souls of the laboring population, at this period, would be a most destructive blow to both Northern and Southern United States. The Irish, the Germans, and the negroes, are the life and soul of America; probably the Germans are not so much so in proportion to the other two races, as they are more given to store keeping, &c., which are unproductive occupations, and only multiply more drones in the hive of the laboring bees, and impose more heavy burdens on them. Take away the negroes, and all the agricultural, and all public works, would fall heavily on the Irish, and Irish descendants almost exclusively. And as the population of Ireland is reduced two fifths from what, under anything like favorable circumstances, it should be; and immigration from that quarter is now from the stock, and not from the excess of its population: in a short time it must be checked.

The Spaniards expelled the Moors, and dealt a heavy blow to their nation's progress. The loss of a body of

* At an increase of 28 per cent., in 150 years, the slaves would be 128,126,630, more than 1 to 5 of whites.

at least two thirds of the agricultural laborers of the United States, would be so destructive to her interest, that no American in heart could countenance it. It would be destructive to agriculture and commerce at the same time.

However, the other evil is to be guarded against, viz.: the overruning America with a negro race. The increase per cent. of the slaves of the Southern States has been extreme, compared to that of whites; who, with all the extraordinary immigration in the last 20 years, have not greatly advanced their per cent. over that of the negroes. In 1830, the slaves were as one to five of the whites; in 1850 they are as one to six nearly. During all that time, the amount of immigration of whites has been 1,855,643, and that 1,855,613 in the prime of life; and adding, by their capability to increase and multiply, as well as by their own numbers, a per centage to the white population, that must be far greater than the negro increase. Therefore, their increase could not be less than 30 per cent., which would lower the increase of the permanently settled whites to about 28¾ per cent. average for the last two decades.

Therefore, the introduction of slaves, day after day, into Brazils, with the productive powers to propagate their kind, there is every prospect of America becoming seriously inconvenienced in some not very distant time. In 150 years hence the number of negroes from the present stock of 8,179,589 will be 70,000,000, at the rate of twenty-three per cent. each decade, or at the rate of twenty-three per cent. 128,126,630. The present amount of the race is computed at 55,000,000 in Africa and all other places. If to the increase of the above amount of the now stock of the Southern States there be added that of the Brazilian slaves, and of South America, and their increase, then there is an almost certainty of converting America into an Africa.

The introduction of more slaves into Brazils, or into any portion of America, will magnify the evil; and it behooves Americans at once to resolve upon a determined line of action, to suppress all further traffic with Africa for slaves. In fact, to declare war in the most decided manner against the further importation of the negro race. England is really and truly now spending some 5,000,000 of dollars annually for no other interest than that of the white race of America in the future day.

At the same time, it must be well understood that there is no one object, no one virtue, that cannot be carried to excess. At present the emancipation of the slaves in the Southern States would be a death blow to both Northern and Southern States, and to the injury of England herself. All true Americans must join in a practical and a humane object, for the good of both races and their future well-being; and wisely and generously and in brotherhood devise the means of a future total separation of the black from the white, and yet to advance the prosperity of both.

The facility to effect that final separation, that advancement of the two interests, is as easy as it is for America's best scholar to read his A B C. I say easy; there would be labor, as there is in all things, but the profits arising from that labor would be so great and extensive that it would be like travelling in a spring-cushioned carriage, of the latest improvement, on a well Macadamized road.

There never was an opportunity equal to that now at America's disposal. There is a wide field, a boundless area in Africa for the multiplication of the human race. It has been, from the first dawn of man, and is now, the natural home of the negroes. They are there, in the year 1851, unimproved and savage. To improve and advance them in civilization would be to work out the merciful designs of an all-wise Providence, and it would be to place the poor negroes high in the human family. To achieve so benevolent an object would be one of the most glorious acts that ever has been recorded in history; one of the most noble and humane for our race and age to work out. Improve and instruct the father, and let him bring up his child from his tender age in the way of thinking, and that child will be susceptible of greater improvement than his father; and that instruction continued from father to son, each successive generation would increase in mental powers. The more exercise the body goes through the more it is capable of enduring; and it is the same with the mind, it increases in vigor. To say the negro is not capable of improvement is quite erroneous; the condition of the slaves in the Southern States explodes the absurd charge, and it is only a dogma put forth by atheists. The white races would descend to the level of the negro capacity if left for ages without education, and circumscribed in their intercourse. Take a man from the coal mines of England and place him beside the negro, and where is the difference in intellect or morals? and yet the miner may be the son of a father who had some education, and he himself had some intercourse with intelligent beings. He sees railroads, machinery, &c., and hears of great goings on in the world, yet he is as low down in the scale of understanding as any negro of the South.

The salutary effects of improving the negro race would be of far greater interest to all than if the whole negro race had been converted into slaves. What would be the consequence of having an enlightened and industrious people in Africa? Open out that unknown continent, to make it productive of all the blessings it is so capable of bestowing on the whole race of human beings. America would then hold a proud position, placed between Europe and Africa on the one hand, and China on the other, carrying on commerce with them all.

An extensive preparation has been for a long period in operation for such an undertaking; and, if applied, an extensive progress would be made by making a commencement. There are some thousands of freed slaves; and the Southern people, of their own good will, have freed, and will continue to free thousands, year after year. All the slaves have been brought up in the same way as the whites—some are well acquainted with agriculture—some have been engaged on ships, on steamers and railroads—some are good carpenters, smiths, &c.

These people, led back to Africa under intelligent leaders, and established as a colony, would do much to improve their race. They would at least be able to extend their influence, and by the aid of America and of continued reinforcements of manumitted slaves from this country, would soon establish a new kingdom and a new state of things. Such a colony, directed by a few able whites, would find it an easy matter to progress. Often in India a few enterprising men establish themselves in the most powerful kingdoms; for instance, the Pindaree chiefs, &c.; and even a few Europeans have raised themselves to thrones in India, without even a cent to commence on. We read in Reynolds's History of the Indies, what a few hundreds of Portuguese have done in East India; of an intrepid leader with a small band defeating an army of thousands, and at a time, too, when the mode of warfare in Europe was conducted without the knowledge of the science of the present day. We know what the Spaniards achieved in America; we know what a mighty kingdom a few merchants have raised in East India; we know how often and often colonies of a few scores of whites have located themselves in these States, and succeeded in progressing against thousands of Indians, and all the other obstacles that opposed them. In no one of these cases I have mentioned, was there the same advantages held out, as there is now, by the collection of these freed negroes, to achieve the same revolutions in Africa. No one can say the project is mere speculation. Had the band of active, energetic men, who went to Cuba, collected a few thousands of negroes, and placed themselves at their head, and set themselves down in Africa, who can doubt that the result would be rapid strides in progressive improvement, not by war, but as Mr. Penn did in Pennsylvania. Neither the Spaniards, Portuguese, Dutch, nor English, ever had the same facilities at their command; and when we compare the science and improvements of the present with those of two centuries ago, say 1650; and, on the other hand, consider that the negroes in Africa are now in as benighted, yea, in a more benighted state than the American Indians, no comparison could be made between the East India people and the negroes. There would be but little exertions needed to gain a complete control over the whole race. *Some* 30,000 *whites in East India govern some* 200,000,000 *of Indians.*

To establish such a colony in Africa would be a far simpler thing to effect than to conquer Cuba, and it would be a means of gradually relieving this country of the negroes.

To emancipate the slaves and leave them in these States would be a step full of the greatest danger, and the most evil consequences would be certain to follow. In America, being freed, there would be no one, as there would be in a colony, to guide and direct them as a body. Therefore the negroes, except in some cases, are not prepared for freedom. They have passed from generation to generation without the habit of thinking, and thought is the mainspring to action. The labor of slaves in the Southern States is but a kind of healthy exercise, as their great increase in numbers show—and they are altogether taken more effectually care of by their owners, than are the owner's own children ; for affection of the parent is often the ruin of the child, but the negro is ruled by judgment tempered with kindness, and a great deal of leniency. The slaveholder knows that to preserve the negro strong and vigorous in body is for his interest ; that ill-treatment or connivance at dissipation would be detrimental to himself ; a sick negro is a burden, a healthy one is wealth to his owner. Therefore these considerations, and the naturally kind and generous feeling of the people of the South, have rendered the negroes the most contented and cheerful people that I have met with in my extensive travels. The abolitionist who would do aught to interrupt that contented condition without being prepared to better it by some practical scheme, is but a selfish demagogue, and should be discountenanced by every well-disposed man.

To bring round the immediate emancipation would not be for the interest of the slaves in the South—and it would be an unjustifiable plunder of the Southern people, as the greater portion of them have no property but slaves. It would be depriving them of that property transmitted to them by England itself.* It would leave the father with his family destitute. It would leave the widow and the orphan to famish in want and misery. It would be to make vagabonds and a discontented people of the now cheerful and contented slave; it would be to cut off the right arm of the United States generally; it would be to bring round a deadly warfare between the two races, which would end in the negroes being exterminated, and would for a few years deprive England of one half of her supply of cotton.

Two articles would cease to be cultivated in America with the emancipation of slaves, viz: cotton and rice.

The yearly exports from the whole of the United States, taking the average of the last three years, would be $133,500,000.

* Slavery was introduced as early as 1620 into South Carolina by the English, and at a very recent period the following parties took part in the trade, viz:

From 1st January 1804 to 31st December 1807, the following is the number of slaves imported into South Carolina:

By the English - - -	19,649
By the French - - -	1,078
In American vessels - - -	18,048

So it is seen England can swear like John Bull at slavery, but make money on human flesh when the opportunity serves.

There are frequent discrepancies between the different reports of exports from the United States in different works; in fact there is no country so deficient in statistical information as the United States.

The yearly exports of cotton, taking averages of last three years, as given in the London Times of 13th October, 1851, is at Liverpool prices, $78,772,361
The yearly exports of rice, say 136,282

Total average exports of cotton and rice $78,918,643

I ask Americans, could they afford to strike off $78,918,643 from their agricultural and commercial resources? Nor would the loss stop at that sum even; cotton and rice would be imported at much lower rates by England from other countries than they could be grown here for domestic consumption by free labor. Therefore the value of these articles consumed yearly in the United States would be lost also, or a sum amounting to $9,100,128, making a total sacrifice of $88,018,771.

Emancipation would be the destruction of the United States from one end to the other.

Cotton now stands the planter, with slave labor, 5½ cents per lb.; expense of cultivation, and ginning—slave labor stands him in 21⅔ to 25 cents a day, (as shown in table C.) For white labor, the planter will have to pay 50 cents up to one dollar a day, which will raise the expense of cultivating cotton, as 21⅔ is to 50, or 12⅔ cents per lb. Four cents per lb. above market rates. But beyond this—WHERE IS THE AMOUNT OF LABOR TO COME FROM? England has had sad experience in the West Indies, of the effect of emancipation, which reduced estates from the value of £50,000 to

£4,000 or £5,000. Abolitionists represent that falling off in value to other causes, but they are mistaken; the decline was so rapid that other causes had no time to operate, nor do the English themselves believe that other than emancipation of the slaves had aught to do with the decline of the West Indies, and we see them doing all they can by importing Bengal coolies to supply the place of the slaves who refused to work.

Dismiss 3,000,000 laborers to-morrow, how could an equal number be supplied?

It would be very trying to the endurance of white males to labor in the hot South; white females could not do so. The present number of laborers in America is calculated at one to five of the whole population. I calculate it at one to six for agricultural laborers; here would be nearly two millions of actual laborers redeemed from the necessity to labor, and to get two millions of substitute whites, it would be necessary at least to import some eight or ten millions of whites. For two million of males could not be exclusively found. The whole immigration of twenty years into this country was about 1,855,643—so the laborers necessary could not be obtained; and to ensure any number of laborers at all, wages would go up to a higher rate than it is at present.

What would the slaves do if emancipated? To get the necessary means of life would be an easy matter. They would plant some Indian corn, and probably some ground-nuts and sweet potatoes; that would be the extent of their agriculture, and possibly they might keep some poultry and pigs; but a large portion of the more intelligent would flock into cities. Some would establish little shops and brothels, others would hire themselves as butlers, &c., and a general system of petty theft would be carried on in every city where they would establish themselves. They would become objects of derision and hatred to the white population; and the few who would amass wealth would only be the cause of greater temptation to the rest to flock into cities. However, the more ignorant might remain in the districts and cultivate these articles mentioned; but they never would do so to the extent of their wants; and then thefts would be the only way left to obtain means. The negro is naturally indolent, and as a hired laborer would be employed with reluctance; and as no planter could depend on getting him to labor at the necessary times, the planter would not run the risk of planting where there was no certainty of his being able to save his crop; and the slave's chance to get employment would be but poor. The emancipated negro, under the circumstances, would become a nuisance in these States, and would, in course of time, be cut off from the land. But allowing that negroes would render themselves happy, and that they would take care of themselves; still cotton and rice planting in the South solely depends on the preservation of the present state of things; for raise the scale of labor to the height that it necessarily would be after emancipation, the cost of producing the above articles would be as stated, at lowest, $12\frac{9}{15}$ cents per lb. In some parts it is even now 6 to 7 cts. per lb., with slave labor; therefore, cotton would cease to be planted. Rice, too, would not only be unprofitable, but I fear the constitution of the whites could not endure the unhealthy exhalations from the paddy lands. I doubt even if the negro, who never had been accustomed to rice planting, would not suffer greatly from its bad effects.

America holds the first place in the cotton market, because she is able to produce a cheap article. Raise the price of labor from 25 to 100 cents, or even 50 cents, and the expense of cultivation will be quadrupled, or at lowest calculation doubled. The expense of cotton cultivation is now from 5½ to 7 cents per lb. India can produce her best cottons equal to fair New Orleans, and lay it down in Liverpool for 7 cents per lb.* Raise the expense here to 11 or 14 cents per lb. of cultivation, and you would enhance the price of Indian cotton to that or any other amount of increase for the expense of cultivation; therefore, the following would be the state of things :—

Slave labor at 25 cents a day, expense of cotton cultivation, &c., - -	5½ to 7	cents per lb.
Free labor at 50 cents a day, expense of cultivation, &c., - - -	11 to 14	"
Free labor at 100 cents a day, expense of cultivation, &c., - - -	22 to 28	"
Expense of cultivation and landing East India cotton of fair New Orleans kind at Liverpool 3½d. per lb., or - -	7	"

So that emancipation of the slaves at present would give a premium of 100 to 400 per cent as an advance in price of cotton, to planters in India, in Brazils, and in Egypt. Can any man doubt the result? An advance in price of cotton to 100 per cent., or even 50 per cent. above present prices, would have the effect of stopping half the cotton mills all over the world.

"The failure of the cotton crop in 1846 (Manchester

* See English Price Currents, or Royle on Cotton Cultivation in India.

Guardian, Jan. 23d, 1850, see Royle) as in the very last season, caused a considerable rise in the price of cotton; and it was calculated that in that year an advance in price of 2d. (four cents) a pound required an increased payment by this country of £4,000,000 sterling. In this year the increase in price has caused many spinners and manufacturers of coarse yarns and heavy goods either to stop their mills or to work short time, and, of course, to throw many of their workmen out of full and regular employment. It has been well ascertained that, with high prices of the raw material the present enormous production of cotton manufactories will not, and cannot, be taken off by the markets of the world."

Can any well disposed American desire to deprive America of fully, if not more, than one half of the whole amount of her exports, $78,918,643, and value of domestic consumption to the amount of $9,100,128? And for what purpose would this sacrifice be made every year? Simply to send 3,179,569 of negroes adrift upon the American public, without any provision made for them. It is but just a day or two since 200 famishing emigrants were obliged to return to Liverpool from America, because they could get no work; and Abolitionists, without taking any thought for the future for so many millions of people, wish to see them on one hand deprived of a comfortable home, and at the same time on the other, to deprive these States of one half their means to give employment to any body of people, or in fact to support themselves; in short, to shut up all work, and to put an end to all progress.

But, to go further into the subject in its economical bearings, it is necessary to get a glimpse at the actual expense a slave is to his master. Taking the value of the grain, the animal diet, &c., which he consumes, and his clothing, to be 50 cents a week (which I believe is a very moderate calculation),* then as all slaves cannot work from their birth to the age of 90, there consequently must be a support allowed for the two terms of his existence—viz., childhood and old age, for sickness, and for one day in every seven; for insurance of life, for taxes, and all the risks that is liable to be encountered by his absconding, &c., and for expenses of superintendence.

From the "Census of Charleston" (printed in 1849), an interesting and valuable work, I have taken the following:—

The number of slaves out of 100 who have died during the following periods are, viz.:—

			Died by.
Of every 100 born, there died before	1 year—	21.64	21.64
" " from 1 year to "	5 years—	16.78	38.42
" " from 5 years to "	10 years—	2.79	42.21
" " from 10 years to "	20 years—	7.52	49.73
" " from 20 years to "	30 years—	9.13	58.86
" " from 30 years to "	40 years—	7.94	66.80
" " from 40 years to "	50 years—	8.43	75.23
" " from 50 years to "	60 years—	6.85	82.08
" " from 60 years to "	70 years—	6.20	88.28
" " from 70 years to "	80 years—	4.82	93.00
" " from 80 years to "	90 years—	4.32	97.42
" " from 90 years to "	100 years—	1.76	99.47
Upwards	100 years—	70	99.87

Which, calculated as in table, will show that of the above

* I have seen in the Patent Office a report, that the supplies for a slave in Mississippi is 25 dollars yearly, not including grain and vegetables.

100 negroes, from birth to their death, they labored 518,638 days; the number of days supported, 1,045,732. Therefore, the days there would be work done are as 1 to 2 of the days supported:—

Therefore, to 2 × 50, ...	100 cents per week.
Life assurance, say 400 dollars each, at 3 per cent. per annum,	23 " "
Medical charges, &c., per week, . .	7 cents per week.
Total, ...	130 " "
Or per day, for six days per week, . .	21⅔ cents.

The expense of superintendence, &c., will make it fully 25 cents, or more. There is the risk, too, of slaves running away. Therefore, in the due course of time, as the population increases, and planters can go to the market and get laborers when required, and at more reasonable rates, slaves will be less needed and less valued. There can be not the least doubt that slavery, about which men are now disposed to shed each other's blood, and to exterminate each other, will in a few years assume a different feature; and is it not better to wait with a little patience until labor can be obtained, than to throw the whole country into poverty and confusion, and reduce it to years of bloodshed and extermination, that will inevitably follow and violent measures by abolitionists? Twenty years hence slavery will be a thing very easily dealt with. At present there are but a few leading staples in America; any to enter into their cultivation, it is necessary to retain laborers all the year round. Were there a greater variety of products, it would widen the field for employment, and one crop would come into the barn before the other would be ripe; and thus the intermediate time, say, for instance, between planting cotton and in picking or collecting the pods, could be occupied with other staples. This cannot be the case now, nor never will, while America confines herself to cotton and corn, and pig feeding. There are no people in the world so little employed as the slaves of the south, because of this state of things; and the man who represents them as hard-labored, must be altogether ignorant of the facts, or a designing fomentor of discord.

In all the old countries, India, &c., slavery is but a nominal thing, so much so, that it passed into a term of courtesy and compliment—Ap ka golambundee hy: "Sir, I am your slave." And when the British East Indian Company say they abolished slavery in the East, they did nothing more than pass an act against a term that had no meaning. Among all the higher classes in India to-day, there are slaves—who are so, willingly; who might be more appropriately called hangers-on, because they cannot do better. Slavery is only necessary or useful where labor is scarce. Cheap labor and slavery are incompatible in the same country, except such slavery, the worst of all slavery,—the hard-worked, ill paid, ill treated servants of England. Let the abolitionists go to England, and travel through that country; let them inquire into the state of poor servant-maids, shop-boys, and farm-servants there, return to this land, and take a tour through the Southern States, compare the slave's condition with the servant's, and then ask themselves, before God and their country, which is most deserving of their help. There are many Mrs. Birds, and Mrs. and Mr. Sloanes in England, who beat and starve young girls even to death. Let them, (the abolitionists) look into that most horrible, most disgusting and filthy scene that man ever beheld, or could behold in no other country than England—a cheap lodging-house—thousands of which are in London. A "cheap lodging-house" is a place of rest for the most miserable, miserable in worldly circumstances, miserable in infamy and vice. Poverty and neglect of "the would-be charitable," draw both the innocent and the outcasts of depravity into the same house. The room of a "cheap lodging-house," where there are beds, may contain half-a-dozen or more, depending on the number that can find place. The room is filthy as it can be, the walls are darkened over with bugs and vermin of all kinds. The charge for a night's lodging may vary from one half-penny to two pence per night. Those rooms for one half-penny and a penny, have no beds.

In the evenings the poor, honest, laboring man has to return there when disappointed in getting employment, no matter how virtuous or moral, himself and his famishing wife and little children—they have no other refuge from the inclemency of a cold winter night. There retire the young ruffians of London, with their depraved companions, whom they call "gals," and upon whose prostitution, with that of their own and these young girls' robberies and pick-pocketing, they live. These young wretches, young in years, some of them not more than from twelve to fourteen years of age, yet old in vice, keep two or three "gals," but younger still there are of the most vitiated habits. There are the little children brought up to thieve. There too are the young men and women, wrecks of vice. There too drunkards, and every other species of depravity, retreat. All there meet in that squalid filthy room ; men, women, boys, and girls sleep promiscuously in the same beds, as they may happen to get a place in them ; if not, they stretch themselves under the beds, or any place they can find room. A tub or pail is placed in the centre of the room, to which all resort. The stench is sickening, for the tub remains for days without being emptied. There is heard the blaspheming of the drunkard, and the diabolical language of depravity. The keepers of cheap lodgings are invariably the receivers of stolen goods, and protectors of vice from pursuit of police. It is in such places that virtue also has to seek a refuge. And there the young are exposed to the severest trials, suffering the direst want, hungry and shivering, and without clothes, and no work to be obtained. They hear stories related of the success in pickpocketing and burglary, and all other methods devised by want and dishonesty to obtain means of existence. They hear every good quality jeered at. Death from starvation is before their eyes. Yea, it has already commenced its work, and they see themselves cadaverous, attenuated shadows of what they had been, and without apparel. Salvation of their existence is held out to them by forsaking those religious principles in which they had been brought up. There is no aid, no help, the charitable declaimer against slavery spurns them with rudeness, yea, I may say, with ruffian severity, that could come but from canting hypocrisy. Thousands of thousands of these poor creatures are left to perish. Rome, in its decline, sold its children in the public market places. The poor of London would sell their children and themselves, to save themselves from death and from infamy, if they were allowed the opportunity to do so, and to them the state of slavery would be the greatest boon. Go, abolitionist, and exercise your humanity and your love of freedom in England, release some millions of people from the grasp of hell and starvation, make them slaves to yourselves or others that may be charitably disposed to receive them under protection, and remember, oh remember! they are your own blood, and Christians, redeem them, and God will crown you with blessings. Save their souls. And when the English agents come over here to create political capital by intefering in your domestic institutions, by your example, show them the sphere wherein their sympathies are required. Leave your slaves in the enjoyment of a home and means of support, until you can do better for them. You already see some of the States passing laws prohibiting the freed negro entrance amongst them. Let three millions of people be exposed to want, spurned by most of the States,—what can they do? They would flock to those States that were so zealous for their interests; Boston, New-York, &c., &c., would be inundated with them. And in this city the abolitionist has only to wander off Broadway into Church street and its purlieus, and ask himself, is this the place of the glorious free negro? But let him not wait until dark, for there the freed negro and the infamous reside. Emancipation would have the effect of extending Church-street over one fourth of New-York, and instead of a few thousands now easily controlled from their paucity of numbers, after emancipation there would be some 50,000 or 60,000 at least; their number and want would make them bold and daring, and the municipal taxation would be three-fold as heavy, to provide more police, more prisons, and support for prisoners, and all other expenses necessary—besides, the city would be ever a prey to sickness and plagues arising from the filth of the negro part of the town. Therefore, it is necessary for all Americans to make preparations before they do resolve on emancipation.

THE PEOPLE OF AMERICA ANGLO-SAXONS.

Every one who reads the London *Times*, and the London journals generally, will perceive the flattering unction they lay to their souls, that the Americans are Anglo-Saxons, and the proud boasting of the wide-spread of that race, when, in fact, the race is, and has been, long declining.

From the earliest periods, America has been a refuge for all nations and races of Europe. The Spaniards were the first who made their way to America; then followed the French and the Dutch. The English had not a navigator to start on any unknown seas, and were indebted to a Venetian for exploring parts of America for them.

Spain colonized South America and part of the Southern States, to which she gave the general name of Florida. The French colonized the North, or Canadas; and, until a few years ago, England found it necessary to make use of the French language in all their public documents. The chief claim of England is, that she has possessed herself for a time of America. This is the only claim she can, in fact, make; for the Anglo-Saxon race in America is not one in ten of its inhabitants.

The Spaniards, the Celts, and the French and Germans, are the people of America; and it is pretty certain that the Celts are the people who preponderate in America. The political and religious animosity of the Saxon to the Celt has from the earliest period of the unfortunate entrance of the Saxon on Irish soil as rulers, driven them as exiles to all parts of the Continent of Europe.

In France, in Austria, and in all parts of Spain, are found the Celtic Irish race to-day; and many of the Spanish commercial towns are principally peopled by Irish, or Irish descendants. America was then open to them. At that early period, when the Reformation gave a fresh impetus to emigration from Ireland, the English were little disposed to emigrate.*

Even to-day her people are adverse to leave the country; and few of those who do leave, and who are put down as Anglo-Saxons, are, in truth, really so. Extraordinary circumstances at times elicit the truth, such as the falling off of the population of Ireland since 1840; when the *Times* of London stated that the emigrants to America from England were Irish, who made that country only a stepping-stone to the United States.

Even now, when the spirit of migration seems to be moving all the States of Europe, the Anglo-Saxons respond not; they hang back; and those few who do emigrate are of the middle-classes, principally merchants and shop-keepers. In fact, if the real state of affairs were known, it would be made clear that the Anglo-Saxon race is not at all on the increase. The whole amount of her emigration is not equal to her immigration; and it may be said that some of her most populous cities are Celtinized, if I may so speak. There is no part of England where it is not the general complaint, that the

* The pilgrim fathers may be said to be the first permanent settlers of English, in 1620

Irish lower the price of labor. Liverpool, Manchester, and Birmingham are instances where the very poorest class of Irish have been for a long period making their way, and doing all the hard and humblest labor, and from that position working up to the top of the ladder. The Irish of Manchester are, it may be said, the people of Manchester. In 1830, there were 75,000 of Irish in Liverpool; and what a great number there must have arrived since then, especially as they raised the censuses of Liverpool from 1840 to 1850 to 50 per cent. In London, Irish and Scotch make a very large proportion of its inhabitants; and if all foreigners, and those of foreign descent, be subtracted also, the Anglo-Saxon race would be probably in the minority in that great city. The armies of England, again, are in the majority Irish. The population of the Canadas is, to a very large amount, made up of Irish and French.

We know that there has been little or no inducement to prevail on the lower classes of England to leave their homes; they had always three or four times the wages and opportunities that the Irish had to make a livelihood at home. The Irish, on the contrary, never could get daily employment in Ireland, arising from the embarrassed position of the landed proprietors, many of whom were Englishmen, who never visited the country, and who not unfrequently mortgaged their Irish estates to supply every pressing demand, while their English estates they fostered and improved. The proprietors of the soil were also universally in religion adverse to their tenantry, and were aliens to them. In the majority of cases they were the troopers of Cromwell, &c. These causes forced the poor, unprotected, persecuted Irish from their shores.

They lowered the price of labor in England; they were everywhere in the Englishman's way, and the English hated them, and at the same time feared them. America became the Irishman's home. On her soil, and under her mild laws, the persecuted, despised, maligned Irishman raised himself to the first rank;* and the second generation held their head high in all things connected with American matters. On their arrival in America, they lived in the poorest places, often in places where the infamous resorted; they labored hard, economized, and, with a generous heart, gave their hard-earned savings to their relatives, to enable them to leave a land crushed by oppressions of a profligate and dishonest aristocracy, who were aliens to them and their country.

Hence it was that the poorest Irish were provided the means of emigrating to a happier land and an honest people; while the poor of England had neither the same reasons to flee their home, nor the means to emigrate. Neither have the same class of people the same enterprise that the Irish have; and instances of this may be produced in hundreds of thousands of cases, where Irishmen get as much money as pay their passage to America, and live on the poor, and even unhealthy, diet supplied to them on board ship, and are set down at New York, Philadelphia, Boston, &c., without a cent to get a night's lodging. It is not only men who are so situated, but even poor girls, who never before saw aught of the

* The Irish were the first, or at least amongst the first, to promote the manufacture of cotton cloths from American cotton. As early as 1790, we find that the Irish had been supplying a large portion of the surrounding country by a manufactory established by them near Murray's ferry, in Williamsburg.

world beyond the landmark of their village and parish chapel. This will be better substantiated by the following article from the London Times:

[*From the "London Times," Oct. 7th,* 1851.]

"The Celtic exodus continues to be the marvel of the day. From morning to night, from the arrival of the first trains before day break to the last which reaches in the evening, nothing scarcely is seen along the splendid quays which adorn Dublin, but the never-ending stream of immigrants, flying as if from pestilence, to seek the means of existence, which their own inhospitable land denies to labor; and the modest ambition to live and die beyond the gloomy precincts of the Irish workhouse. Numbers of these adventurers are of the better class of farmers, and appear to lack none of the appliances requisite towards the bettering of their condition at the other side of the Atlantic; a healthy and more than comely progeny, a good supply of the most requisite articles of furniture and clothing, with some small capital to commence operations. The majority, however, have no such advantages to boast of; for a more miserable, sickly looking, and poverty stricken set of creatures it would be impossible to imagine; even hundreds of them—men, women and children—being unprovided with shoes to their feet, and the females with no better covering to their heads than the commonest cotton handkerchiefs in lieu of bonnets, while not one in fifty could lay claim to the luxury of a cloak as a protection against the inclemency of the coming winter. All hardships appear as nothing, so that the one great end may be achieved—

flight from the British shores ; no matter at what risk or with what amount of danger and privation in perspective. Day after day vessels leave this port freighted with human cargoes, without any diminution being perceptible in the throngs of peasantry which swarm the streets in the neighborhood of the quays.

The rush, too, from the South is rather on the increase than otherwise, and is on a far more extensive scale than we in the metropolis have any idea. On Saturday a steamer left Waterford for Liverpool with nearly 400 immigrants on board. The day was intensely severe, but wind and weather, be they what they may, have no terrors for these voluntary exiles. The average number from Waterford alone, since the season set in, is 500 weekly."

These people worked and rose in the scale of society ; and day after day American money flowed into Ireland,* and Irish emigrants sought these shores. America prospered ; and the English now turn round, and are but too happy, even at the expense of truth, to put down, not only the Irish emigrants from English ports as Anglo-Saxons, but to represent the whole of the Irish as British emigrants, by carefully avoiding to note or classify those who do leave the British or Irish shores.

Taking another view of the case : the whole world has heard of the powers of the Irish to propagate their species. Many have been the nostrums put forward by the people

* The amount of money sent to Ireland is enormous, through the Catholic clergy and Agency Houses. "One agent in the city of Cork has acknowledged the receipt of £1,000 in one day from America."—*London Times, Oct. 8th*, 1851.

of England to prevent the over increase, by preventing early marriages, and marriages of poor people. All England echoed, from one end to the other, with declamations against Irish procreative powers, and the misery it entailed on Ireland. If the *London Times*, and leading English statesmen, were correct, what has become of that increase in the population? Let them take a glance at the census for the last 50 years.

	England.	Ireland.			England.	Ireland.
1810	8,331,434	5,191,240				
1811	9,538,827	5,637,856,*	increase	per cent.	$14\frac{1}{2}$	$8\frac{3}{5}$
1821	11,261,437	6,801,827,	"	"	$19\frac{1}{5}$	$20\frac{3}{5}$
1831	13,091,005	7,767,401,	"	"	$16\frac{4}{11}$	$14\frac{1}{5}$
1841	16,995,508	8,175,124,	"	"	$14\frac{7}{13}$	$5\frac{1}{4}$
1851	16,594,275	6,515,794,	"	"	11 decr.	$25\frac{1}{2}$

The increase of 20 per cent. is but a moderate increase. What then has become of that ever dreaded multiplication of the Irish? It is to be found in England—in Birmingham, Liverpool, Manchester, London, and throughout the whole of England; in Glasgow, and in all the towns in the British empire.† Then, what is become of the Anglo-Saxon race? Their increase is below par, and very much so; and if the Irish be substracted, if the Scotch also be deducted, and foreigners, then, let me ask, what has become of the increase of the Anglo-Saxon race? The Irish can be found. There are millions of them and their descendants in England. They are to be found in British America, in the United States: where

* 5,937,856, census of 1812.

† 7,000,000 are computed to be in America by Mr. Robinson.

is the English population? It is to be regretted that the United States authorities have been so neglectful in keeping returns of all classes of people immigrating into this country. There are hardly any returns, and what there are, are vague and unsatisfactory.

The arrivals of immigrants at New York for 1848, were—

Irish, . . .	98,061
Germans, . . .	51,976
English, . . .	23,062
Making with immigrants from all other places, . . .	189,176

The arrivals at New York for 1849—

Irish, . . .	112,691
Germans, . . .	55,700
English, . . .	28,320

The arrivals at New York for 1850—

Irish, . . .	116,583
Germans, . . .	45,404
English, . . .	28,131

These tables will show to which race the United States are indebted for their great increase in population, and will show how little it is in keeping with the real facts of the case, to represent the people of the United States as the Anglo-Saxon race, inasmuch as they are not one in ten.

Again, the whole amount of emigrants from Great Britain and Ireland, during the same years, to all places, was as follows:—

			British Poss.	U. States.	Australia, &c.
1848,	248,089	of these	31,065	214,283	28,791
1849,	299,489	"	41,365	219,450	38,681
1850,	280,436	"	32,965	223,078	24,810
Total,	828,014		105,395	656,811	92,282

The number of emigrants to the British possessions in America, bear, if not a greater proportion, at least an equal one, of Irish emigrants, as the numbers of Irish to New York does. The emigrants to Australia of farmers from Ireland, preponderate equally over the British and Scotch.

It is found that in three years alone, out of 25 years of returns, there are of Irish to New York, .	337,335
Say the Irish were in the same ratio to the British Possessions. English to New York, 79,513; Irish, 337,335; or say $\frac{79}{416}$ were English, $\frac{337}{416}$ Irish,*	85,383
Say the total emigration from Great Britain and Ireland for the prior 22 years, to be as per returns, 1,794,181 $\frac{79}{416}$, of which were English and Scotch, $\frac{337}{416}$ Irish, or	1,453,700
Total Irish to all America for 25 years, . .	1,876,413
Total emigrants from Great Britain and Ireland to America, ...	2,325,026
Total English, Scotch and Welsh emigrants, in 25 years, to America, ...	455,616
British and Irish emigrants to all parts of the world 241,007. Say there was a less ratio of Irish, or $\frac{1}{3}$ English. Total English, Scotch and Welsh to all other parts in 25 years, ...	80,335

* Census of 1844 shows the population of Lower Canada at 690,782 (making an increase of 178,863 since 1832) of whom 524,307 are of French origin, 11,895 natives of England, 43,892 of Ireland, 13,393 of Scotland, 85,660 natives of Canada, of British and Irish extraction, and 1329 natives of Europe.

Total English, Scotch and Welsh to all parts for 25 years, ...	535,951
Total Irish to America for 25 years, ...	1,876,413
Total Irish to all other parts, ...	160,672
Total Irish to all other parts for 25 years, not including those to England, ...	2,037,085

Let us return to Ireland, and philosophically inquire what amount of the inhabitants did leave their homes, and if that number or more did really emigrate where they are to be found.

It would be well for poor humanity that there were no classification of the human family, and that we could look upon each other as the children from the loins of one father, and from the womb of one mother, and that we could act so to each other as affectionate children of our first progenitor. But alas! Cain slew Abel, and since then, nationally as individually, we are doing little more than slaying each other. But if the poor Irish strive to rise even to the scale of equality, or even ask justice, England strikes them low. And when under the wise and humane Government of this great and glorious Republic Irishmen raise themselves, and act the part of good, honest, and industrious citizens, England endeavors to make it appear that they are Anglo-Saxons. But truth before vanity, facts before prejudice. Let us kindle the fire for the weird beldames—national hatred and religious intolerance—let us in all cases have the truth. If the Irish have greater procreative powers than the English, let us know the fact, and ascertain the causes, because two agents tend to the multiplication of the human race—

health of the body, and temperance alias morality. If the truth be disguised, the preacher or the philosopher cannot howl vice from behind the cloak of falsehood, and show to man its debasing, exterminating effects. Therefore, without prejudice, for truth's sake, let us consider this matter.

The English Commissioners' Report represent only 2,371,600 emigrants as having left Great Britain and Ireland during twenty-five years for America, and 251,017 for all other parts of the world, making together, emigrants to all parts from Great Britain and Ireland, 2,622,617.

Now, as stated, we know England grumbled like a bear with a sore head, to use a common phrase, at the vast increase of the Irish.

The population of Ireland was, in 1811, computed at 5,637,856.* During ten years, from 1811 to 1821, the increase was $20\frac{3}{5}$ per cent. England increased in that time $19\frac{1}{5}$ per cent. Therefore the Irish must increase more than $20\frac{3}{5}$ per cent. Religious intolerance, poverty and enterprise drove many from Ireland to all parts of the world. Say her increase was 25 per cent.; England that ten years increased largely beyond her usual rate prior to that date as well as subsequently, viz: to $19\frac{1}{5}$ per cent. which could only be owing to immigration from Ireland—and during that time the emigrants from Ireland must be as follows: Calculating the increase of Irish at only 25 per cent., (instead of 35 per cent.,)†

* In 1812 it was by census 5,937,856.

† London journals make it 35 per cent. See Merchants' Magazine of New York.

4^5 per cent. emigrated—or say from 1811 to 1821, Irish emigrants, 251,965.

In 1821, the population in Ireland was 6,801,627—and in 1831 was 7,767,401, showing an increase of $14\frac{1}{5}$ per cent., allowing for the whole increase only 20 per cent. instead of the statement that it had been 35 per cent., the ten years prior, this would show that $5\frac{4}{5}$ per cent. emigrated, or 414,496.

For 1831, the population in Ireland was 7,767,401—and in 1841, 8,175,124—showing an increase of $5\frac{1}{4}$ per cent. only—calculating the natural increase at 20 per cent. it would show that $14\frac{3}{4}$ emigrated, or 1,145,697.

For 1841, the population in Ireland was 8,175,124—and in 1851, 6,515,794—showing a decrease of about 20 per cent.—calculating the increase as usual at 20 per cent. up to 1845 prior to the famine, the population in that year would be 8,992,636, in 1850, of that number only 6,515,794 remained. Therefore more than 2,476,842 souls disappeared from Ireland (there must be some births in '46, '47 and '48) then in 1849 and '50 there must be an increase say at the rate of 2 per cent. for two years—now in this decade 1,684,892 emigrated from Great Britain and Ireland, of that number there has been returned as Irish 1,100,000; and of the 584,471 the London Times says "many thousands were Irish;" which must be the case, as it is well known the majority of the emigrants from England are Irish and Irish descendants.

Therefore, say of the 1,684,892, there were of Irish 1,263,169—this number would be in proportion to the ratio of Irish and to the English who emigrated for the last years to New York. Well, beyond this number, we know Liverpool increased 50 per cent. in the last ten years (the increase of London is on the average 25 per cent. yearly for the last forty years, but in the last decade it was only $17\frac{4}{9}$ per cent) and say 5 per cent. of that were immigrants from foreign or from the country districts of England, it would give $12\frac{4}{9}$ per cent.* Therefore Liverpool at the same rate would leave $37\frac{5}{9}$ per cent. of immigration, and of that at least 33 per cent. was from Ireland, or 86,805 souls. Say therefore of the 1,684,898 emigrants from Great Britain and Ireland, that there were of them from Great Britain as 163 is to 680 Irish, or Irish emigrants, from 1840 to '50, 1,281,015. Irish emigrants to Liverpool, 86,805. Say Irish emigrants to all other parts of England at least two to one for those who remained in Liverpool, say 173,610. Total emigrants from Ireland from 1840 to 1850, 1,541,430.

There are pretty certain data for this calculation, and it shows that there were beyond the usual proportionate number of casualties, 935,412 souls destroyed by famine and its concomitant diseases.

From the foregoing calculation it would stand thus :

That from 1811 to '21 Irish emigrants were -	251,965
" 1821 to '31 " " -	414,496
" 1831 to '41 " " - -	1,145,697
" 1841 to '51 " " -	1,541,430
Total emigration from Ireland for forty years -	3,353,588
Say for twenty-five years - - -	2,901,573
Total emigration from Great Britain and Ireland -	2,622,617

Showing that England received from Ireland more im-

* The whole increase of England, not including Scotland and Wales, was but 11 per cent.

migrants than all that emigrated from her of English, Scotch and Welsh, to the amount of 278,956.

So that the Irish, at the same time they are computed at 7,000,000 in America, are also displacing the Anglo-Saxon race in England.

We can see, that from 1830 to 1840, emigrants from Great Britain, Scotland, and Wales, as per Commissioners' Report, were 140,830.

From 1840 to '50, the "London Times" states, in a vague manner, viz., of 1,684,892 that emigrated, 1,100,000 emigrated from Ireland alone, and there is no doubt that, of the remaining 500,000, many thousands were Irish. Say the "Times" meant by many thousands some 100,000; but the emigration to New-York shows the Irish were as 337 to 79, or very probably more. English were $\frac{79}{416}$, of 1,684,892, leaving of English and Scotch, &c., say, 320,000.

Say, for the five years prior to 1830, total emigrants from the United Kingdom were 177,991—say one fourth were from Great Britain, 44,500.

Total emigrants from Great Britain for twenty-five years, 505,330.

It is well known that the Scotch are a more enterprising race than the English, therefore, deduct one fourth for that branch of the Gaelic or Celtic race of Scotland—126,440.

Total emigrants for England and Wales in twenty-five years, 378,890. Therefore,—

Irish emigrants, . . .	3,353,588
Scotch, ...	126,440
Total, Scotch and Irish, . .	3,480,028
Against English and Welsh, . .	378,890

Not as one is to eight of the Celtic race.

We see by returns of immigrants into New-York, for 1848, '49, and '50, the number was as follows:

		To New-York.	To rest of America.
1848—	Irish,	98,061	62,995
	Germans,	51,973	"
	English,	23,062	20,000
	From all other places,	16,080	
1849—	Irish,	112,691	40,360
	Germans,	55,700	
	English,	28,321	13,563
	Scotch,	8,470	
	Welsh,	1,782	
1850—	Irish,	116,583	86,307
	Germans,	45,404	"
	English,	28,131	26,727
	Scotch,	6,771	
	Welsh,	1,520	

This will fully show the claims of different parties to call Americans after themselves. Surely the claim of England is preposterous.

The foregoing will show, that the emigration from Ireland to England in the last twenty-five years was 1,058,298; besides the Scotch and other people from the Continent of Europe.

And if to this number of immigrants be added those of the Irish race in England prior to 1825, and their descendants, it would be found the Anglo-Saxon race will be seriously reduced.

I will further enter into this matter, and show, by following data from the best available authorities, the different periods of the first settlements in America; of the Dutch, English, and French in these States and North America.

1494. Henry VII. granted a commission to John Cabot, a Venetian, and his son, to navigate all parts of the ocean.

John Cabot discovered Newfoundland.

1497. John Cabot died. His son made Labrador.

1498. Florida was discovered.

1502. Young Cabot entered the Gulf of St. Lawrence.

1518. Martin Frobisher sailed to discover a N. W. passage, and reached Labrador.

1534. Jacques Cartier, from France, made the Gulf of St. Lawrence.

1535. Jacques Cartier sailed up the river St. Lawrence, and wintered in Canada.

From Cabot's time to 1579, England neglected her discoveries, except that a few fishermen remained on the coast of Newfoundland.

1579. Sir Humphrey Gilbert made a voyage, but returned, himself and his people.

1580. He made a second attempt, and he and all his people perished.

1585. First English settlement under Sir Richard Grenville was made at Roanoke, many of whom perished, and the remainder returned to England. To find gold was their object.

1587. Sir R. Grenville set out a second time, and settled fifty men in the deserted settlement.

Sir Walter Raleigh sent out Governor White, who, on his arrival in Virginia, found Grenville's colony destroyed, either by famine or by savages. Gov. White left 175 men.

1590. Grenville returned to Virginia, and found the whole colony exterminated.

1602. There was no European in all America. (Mr. Macgregor must have meant no Englishman, as there were thousands of Spaniards and others from the Continent in South America.)

1604. De Monts sailed from St. Malos, and established a small colony in North America.

1607. The London Company sent three vessels, accompanied by Mr. Percy, who formed the first permanent establishment on James River, with 104 persons.

The Plymouth Company sent two ships to N. Virginia, with one hundred planters; forty-five only remained.

The settlement at James Town was settled in April; was greatly reduced by Sept. following, owing to their bad conduct to the Indians; but were reinforced by 120 persons, principally gold seekers and refiners, and ruined gentlemen.

1609. Lord de la Warre was appointed Governor of the colony, and was sent out with 500 emigrants in nine ships. This colony turned cannibals, eating themselves and Indians, whom they shot, and were reduced to sixty, and finally the establishment was given up.

Same year one hundred planters, under Henry and Raleigh Pophams, were sent out to Sagadahoc, at the mouth of the Kennebec. Most of them died, and the colony was given up.

This will sufficiently show that the Anglo-Saxons are

not people suited to colonization. Australia was colonized by Irish transported there for crimes, or imputed crimes. "The hatred of arbitrary power, in political or religious form, was certainly the predominant cause of the emigration that peopled Anglo-America. Its rapid settlement was caused in a much greater degree by the persecution and disabilities which drove Puritans to New-England, Quakers to Pennsylvania, and Catholics to Maryland, than by the spirit of adventure."—MACGREGOR.

Therefore, too, all know the unfortunate persecution against the Irish, from the commencement of the Reformation.

1614. The Dutch colonized on the banks of the Hudson.

1620. The Pilgrim Fathers established themselves in Massachusetts Bay.

1606. The French established themselves in Nova Scotia.

1608. The French established themselves in Canada.

1628. The next new establishment was made by John Endicott and his wife.

" A settlement was formed in Carolina.

1633. Lord Baltimore commenced colonizing Maryland, having some time prior established the colony of Avalon and Ferryland.

1635. A colony was established in Rhode Island.

1664. Do. do. in New-Jersey.

1669. Do. do. in South Carolina.

Slavery was introduced, it is said, in 1620, in South Carolina.

1682. Wm. Penn arrived in Pennsylvania.

THE UNITED STATES' PLACE IN AMERICA.

The population of all America may be from 55,000,000 to 60,000,000; that of the United States is now upwards of 23,000,000. The States are central, local, and self-governed.

The United States' area is	-	2,081,759,000	acres.
The area of the whole of North America	is	4,736,000,000	"
" " " South America	"	4,160,000,000	"
Population of the British Possessions,	-	2,900,000	
" United States,	-	23,267,500	
" South America,	-	33,000,000	

The United States will have to do with both North and South. They cannot mark their boundaries and say—Thus far we go, and no farther; such resolve could not be adhered to. The Canadas, or British America, are under a foreign government controlling French and Irish, two races equally hostile to England. It is against the nature of things that that government can long exist. The South, it may be said, has only governments that would suit the sixteenth century, or the Nipon Islands in the Japanese Sea; and simply because that state of things cannot last, the United States cannot stand still. Progress must be the order of the day,—unless designing knavish demagogues divide and subdivide this Union into miserable paltry governments, when, in such events, the debris of these glorious States will become the contempt and prey of vagabonds and pirates. Disunionists and abolitionists are the poisonous reptiles that lie hid in the grass; both must be exterminated, must be rooted out, and this great continent become one family governing themselves by Christian doctrines. Love God above all things, and your neighbor as yourselves. Agriculturists must manage their affairs, and communicate with each other through agricultural societies—mechanics through their societies, and merchants through their chambers. And the necessary union and dependence of each society on the other must be freely acknowledged. We all know the first rudiment of prosperity is the fecundity of the soil; THAT IS THE ROOT—and if insecurity, arising from want of due protection, or from anarchy arising from agitators, or invasion of barbarians, interrupts its taking deep hold on a mighty continent like that of America, the whole fabric of society must fall to the ground. EVERY MAN MUST GOVERN HIMSELF, AND MUST ON ALL OCCASIONS BE PREPARED TO MAKE A GENEROUS AND NOBLE SACRIFICE TO THE WELL-BEING OF SOCIETY. If not, a government will govern him with a despotic sway, and he will be but a serf—biting and foaming in the irons that bind and goad him. Be not mistaken, Americans. God rules all; and the man that accepts not him as his God and his Ruler, will be a servant of servants. Those writers, who have their experience or knowledge from the class of books they read, will write only from that class of books—and are imbued only with their principles, except where a holy reverence animates them. And as such, a Gibbon, a Bolingbroke, a Hume, and others, may sneer at great truths; but the men who would go over this globe of ours must cast aside their miserable theories, and bow to an Almighty Benevolent Power, whose designs we ourselves frustrate. "Give us a king," cried the Jews—and God said, "Samuel, give them a king, for they have not rejected thee, but me." And—"This shall be the right of the king. * * He shall take your sons and put them in his chariots, and will make them his horsemen, and his running footmen to run before his chariot; and take your fields, and your vineyards, and your best olive-yards, and give them to his servants. Moreover, he will take the tenth of your corn, and the revenue of your vineyards to give to his eunuchs and servants; your servants also, and your handmaids, and your goodliest young men, and your asses, he will take away and put them to his work." How truly this is the case, the traveller may see; there is no one item of that prophecy that has not been fulfilled. All kings and governments are evils inflicted on us for our want of resolve to govern ourselves individually; and who that considers these States divided into four or five governments, but will acknowledge that each government would be a curse, as above prophesied? for each government would necessarily be separate and divided in interest, and all sources would be exhausted to give them strength. The weakness of the one would place it at the mercy of the other, and jealousy and strife would exterminate a people now too happy and prosperous to bend their knee to the Author of their being, and acknowledge with thankfulness their happy condition. If men will be considerate to each other, and will consider the general good, and live in unity and harmony with each other, the movement of America must be onward, and the government and people must prepare for a state of things hardly to be conceived, and never equalled in the pages of history—a united people, embracing an extent of surface unparalleled; even now, with the exception of Russia, it is greater than that possessed by any existing government. In the Western States rapid progress will be made—a trade unparalleled with China and Australia, &c., will spring up—while the East will traffic with Europe, Africa, &c. Therefore let the United States remember she has but a handful of people, and not sufficient means profitably to employ even them. Let her remember that her cities take up nearly, if not fully three millions of the twenty-three millions of her population; that she has barely twenty millions of a productive population—I mean by productive, those who raise commodities for consumption and for export, and manufacturers who enhance the value of such commodities.

It has been shown that there is an area of 89½ acres to each individual, and if some 2,500,000 citizens be deducted from the 23,267,500, there would be but 20,767,500; and say of that number there be one in every six an able-bodied agricultural laborer, it leaves only some 3,462,000* to till an area of 2,081,759,000 acres, or one man to till and house his produce to every 600 acres. If this be the state of things, can it be said that the United States are preparing for the great events before them? Can one man cultivate 600 acres of land, and make it productive? Yet however absurd this question is, there are men who say we have too many immigrants, even while the whole area of the United States is, I may say, lying one uninterrupted wilderness before their eyes.

* There may be a greater number by the census, for all I know—but the census does not allow for casualties.

Turn up the soil of that two billions eighty-two millions of acres; render it productive and useful; dig out its rich treasure of precious ores; build up your cities; people your wilderness, and then Americans will cease to be elated at their success in building a little yacht, or at their improvements in a revolving pistol; or smile with complacency when a London newspaper deigns to flatter. But what are the means of extending cultivation?

Cotton in twelve years has declined in price 30 per cent. yearly.

Rice in quantity and quality, in nine years, 15 per cent.

Tobacco has declined in quantity the last five years, $2\frac{2}{3}$ per cent.

Bread-stuffs are returning to the same amount of exports that they were prior to the failure of the potato, &c., in Europe.

Sugar cultivation is not advancing. Brazil and Cuba will retain that trade in conjunction with East India and the Mauritius.

There is, under present prices of labor in America, no possibility of extending these articles; and except tobacco, probably they will all greatly decline. We see that India can produce good cotton, and land it in Liverpool at $3\frac{1}{2}$ pence, or 7 cents the pound. In America it cannot be cultivated under $5\frac{1}{2}$ cents per lb. We see not only that they are capable of producing good cotton in India, and at that low figure, but that one of the American planters has given up his employment with the East India Company, and is become a planter on his own account; a thing in itself trifling, but in the symptoms it exhibits, full of significant meaning. England and America have been dearest friends in cotton matters; unfortunately there seems to be, and actually is, (for I have seen both sides of the subject,) an uneasiness exhibited by both parties to be independent of each other. This is discouraging, and it is melancholy. Americans suppose that the Continent will be able to give them better terms for their cotton. This is erroneous; the Continent of Europe has not the same means either in machinery, shipping, or foreign commerce. Therefore it cannot, and it is not able to afford the same advantages that England can, to the seller of the raw material, or the purchasers of the manufactured cloth.

Americans are disposed to shake hands with the weaker party and shake their heads at the stronger; while England, on the other hand, is straining every nerve to render herself independent of America. This may be well in its effect on the whole, but it will sever that great tie that binds the two countries together.

But America must turn her soul from cotton; it can be no farther a national staple; its time has passed; the country has outgrown it, and henceforth it ceases to be what it has been; other staples must be called in. But no one staple of itself ever can be to America what cotton has been. It will require a multiplicity of them, and from time to time she must add to them, with the growth of her people.

America could never have borne the decay of her principal staple for the last ten years, of 125,000,000 dollars, compared with the former five years, were it not for California, and the impulse given to exports in breadstuffs by the failure of the potato crop, which, from 1825 to 1845, varied from 11,634,000 to $16,000,000.

In 1846 the exports went up to				$27,700,000
1847	"	"		68,700,000
1848	"	"		37,473,000
1849	"	"		39,155,000
1850	"	"	only	26,000,900

So it is seen these exports are returning to the old standard; and, with a decline in every article, can Americans stand and look on in apathetic indifference, and, like a Hindoo or Mussulman, when calamities press, instead of exertions, envelope themselves in the fumes of tobacco?

Mr. Walker, late Secretary of the Treasury, had better hopes for his country some years ago; and he estimated the exports according to what he supposed from antecedents, for the past years, more than treble of that which they are, viz.:—

1846, value of exports were			$101,718,042			
1847,	"	"	"	150,574,844		
1848,	"	"	"	132,934,121	estimated	$222,898,350
1849,	"	"	"	132,666,995	"	329,959,983
1850,	"	"	"	134,900,265	"	488,445,056

These are grave subjects for Americans to consider deeply, and to turn their undivided attention to mend, instead of those horrible politics. I have been in these States for nearly eight months, and actually I have heard nothing but politics. No other nation on earth, not even England, with all her morbid intolerance in religious matters, has suffered an equal drain on her mental exertions on this head, as the people of the United States. There are few nations that have an equal number (comparatively) of newspapers and journals; and the whole of them, without an exception that I have been able to meet with, are political. I have been told by an intelligent gentleman, who had been through some of the districts of these States, that he had seen curious instances of the injury this state of things inflicted; and one in particular he mentioned, was that of a fiery politican, with whom he spent a day. The man was the owner of upward of twenty-five cows, but, from bad management, he could not get a drop of milk for his breakfast; yet that man fancied all his evils sprang from government.

It is absolutely necessary to devise some means to ensure the prosperity of America. It was only this day I read of the return to Liverpool from these States of two hundred laborers, and some fifty more who desired to get a passage on the same ship. Alas! that this should be the case. Upon what stands the whole fabric of society? Is it on commerce? is it on manufactures? is it on politics? Who gave to commerce articles to trade on? who gave the manufacturer raw material to work? who is it that maintains that class of citizens who are such stumbling-blocks to civil progress—your scheming politicians? It is that poor man who is to be met with in the field, with his coat lying on the mould, his sleeves tucked up, an old hat on his head, strong, unpolished shoes on his feet, the plough, or the hoe, or the sickle in his hand, the perspiration standing in large drops on his brow, or streaming down his cheek.

He it is who supports commerce—he it is who gives employment to the ship-owner, and all his train of shipwrights, carpenters, coopers, smiths, weavers, sailors, sail and rope makers—he it is who supports the physician, the lawyer, the clergyman, and the legislator—he it is who maintains the military and the navy. Two hundred of these friends of society have left these shores to return to the land from which they came. They came here, one solitary hope animating them, that they could earn, by their labor, an honest and humble subsistence. They came, young, active, in the prime of life. Their country had had the expense of supporting them from their childhood; and when they came to these shores to give their services to the country, they were rejected, because this country will not study its own interests, or will not know them.

There is a melancholy fact made public by the return of those 200 men. It declares that Americans are in that state, that they cannot find means to employ a few hands on the wide area of 2,810,000,000 of acres of land.

Here, politicians, is a subject for your consideration, and a lesson for you to cease your political harangues and scribbling, that destroy all sober and useful thought, and lead men to become frenzied enthusiasts, unfit for any good purpose; harangues and scribbling that banish all national spirit, and sap the foundation of society; that cause men to devote all their energies, bodily and mentally, to support this or that opinion about the government of interests, which, through their own neglect, are allowed to wither away, and fall in ruins under their feet.

Consider the lesson the departure of these two hundred laborers reads to you to-day. Oh! citizens of America, it speaks, in melancholy accents, the sad truths, the very serious position in which you stand. Population increasing at an unparalleled rate, and exports and all your leading staples declining. These 200 men were poor; they would live wholly on the produce of the field; they would require but plain clothing, and of American manufacture; fifty cents a week supports a slave, that would support them, or nearly so—and the labor in the field of these 200 men would support 1,200 American citizens; by allowing them to retire from these States, you so far cripple your own power. You see, in the very fact of their deportation, that the rusts and damps of your indifference to the fundamental interest are corroding the main-spring of prosperity. You see yourselves retrograding 200 steps backwards. It has pleased God to place on the shoulders of other nations the expense of raising laborers for you, and, by a want of good management, or rather by gross neglect, you do not avail yourselves of the blessings held out to you.

I will very briefly recapitulate those staples, of the introduction of which I will respectfully urge the practicability and necessity. First is tea.

I have treated in the foregoing pages of the quantity consumed. There can be hardly less in tea and in substitutes of other leaves for tea than 1,140,000,000 lbs. consumed in China. No one need be doubtful of this. Tea is the usual beverage of the Chinese at and between meals. There must be some 500,000,000 inhabitants in China, unless the people did not increase the last half century. However, say 367,000,000 consume that quantity of tea; we know that 23,000,000 of Americans consume 144,936,892 lbs. of coffee, and 20,000,000 lbs. of tea; therefore, the consumption of so much tea in

China is not wonderful. Say the population of the world to-day is—

Europe, - - - -	250,000,000
Asia, - - - - -	612,000,000
America, - - - -	55,000,000
Total, - - -	917,000,000

There are, of that population, only the Chinese who have their tea at a fair value. 30,000,000 of English and Irish consume 50,000,000 lbs., with a heavy duty of 55 cents per lb. on it, and of a spurious kind. If they could get good tea at a fair price, there would be in Great Britain and Ireland a consumption of 150,000,000 lbs.; therefore, if thirty millions would consume that amount, 250,000,000 of Europeans would consume 600,000,000 lbs. *This can be produced in America for a few cents a pound.*

The United States consume of coffee 145,000,000 lbs. Surely the Government and the people ought to do something to encourage the introduction of the plant, and consider the immense importance of these two simple plants, that do not require annual cultivation. They would not only become an export trade in themselves, greater in amount than all the exports now of America, but would have the political effect of extending the influence of the United States over the bordering States, and over the whole world.

There cannot be the least doubt of indigo becoming one of American staples; and for this plant and the tea plant, I can assure my readers that there can be no doubt of their flourishing in America; and, as I have had long experience in cultivation of both, it is my intention to carry out the cultivation and manufacture of them. The coffee plant can, as stated, be introduced from the same climate as the Southern States, even colder than Georgia. Of the other staples I have said all that need be said; but may simply state, it is my object to introduce seeds or plants, or plants and seeds, of each, and I only trust that the American public will assist me in the object.

I beg respectfully to mention that I am now in America by the advice of the Hon. Abbott Lawrence, and if the reader has time, he may see the correspondence between that gentleman and myself in the Patent Office Report of 1850–51.

The area of all America is 8,896,000,000 acres. The United States already possess one fourth of that surface nearly; and as the people are the only strong party on the whole Continent, Spanish power must give way. Brazil must succumb to the general movement; the Canadas and England are on the eve of separation. Who is there to fill up the places of these foreign powers? This Union, if preserved from Disunionists and Abolitionists, will spread like the young banyan tree over the soil. Each branch will put forth its root, and each root will become a stem of the great parent tree, and all will be peace and happiness. If Europe be agitated, let America sit tranquil and dignified. Her duties lie within her own confines, and within the interests of her commercial intercourse. The opium trade should be suppressed, not by war, but by the moral censure of every honest American.

Great cities and ports must be raised on the West coasts to trade with China, with 500 millions of people, to supply them with grain, timber, &c., and to receive from them such articles as America requires, and cannot produce as cheaply; and it is of paramount importance to America that the opium trade be suppressed.

ERRATA.

Page

19, 1st line, for 4,375,000 read 3,252,684.
30, 12th line, for one-third read one-half
34, 15th line, for American rice read America has shipped rice from Eastern ports to China.
80, 1st line, for North read South.
99, 13th line, for a pretty little bush read pretty little bushes.
137, 7th line, for $36 to $40 read $18 to $20.
" 11th line, for $39.56 to $43.56 read $21.56 to $23.56.
" 13th line, for $53 read $30.
138, 29th line, for one-sixth read one-third.
190, 11th line, for hill read Nile.
194, 28th line, for 8,179,589 read 3,179,589.
207, 25th line, for and read any.
" 28th line, for any read and.
238, 17th line, for 2,810,000,000 read 2,081,000,000.

www.ingramcontent.com/pod-product-compliance
Lightning Source LLC
LaVergne TN
LVHW020312110826
845148LV00017BA/390

* 9 7 8 1 4 2 5 5 2 1 1 7 2 *